SOLENOID ACTUATORS

About the author

Daniele Righetti was born in Perugia, Italy, in 24th July 1985. He graduated in 2010 in Aeronautical Engineering at University of Rome "Sapienza". He works in aerospace industry as designer of electro-mechanical and hydraulic systems. He focused his work on solenoid valves and actuators. This book derives from this experience and collects all studied and ideas developed by the author.

SOLENOID ACTUATORS: Theory and Computational Methods

Daniele Righetti

Youcanprint *Self-Publishing*

Title | Solenoid Actuators: Theory and Computational Methods
Author | Daniele Righetti

ISBN | 978-88-92671-12-6

Cover image by the Author

Youcanprint *Self-Publishing*
Via Roma, 73 - 73039 Tricase (LE) - Italy
www.youcanprint.it
info@youcanprint.it
Facebook: facebook.com/youcanprint.it
Twitter: twitter.com/youcanprintit

To my mother and my father

Contents

Introduction

The aim of this book is to give the reader all the computational tools useful to size and design solenoid actuators. This book discusses only direct current solenoid actuators (Appendix B discusses rectifier principles), that means solenoid valves and many types of actuators used in many different industrial fields, such as automatization, automotive, aerospace, etc.

There are many technical publications that discuss about electromagnetic devices or solenoid actuators and valves, but none of them is able to give formulas and tools in order to analyse solenoid performances or to size the solenoid itself. All of these books give the reader a general view of which are the topics concerning solenoid actuators. Some of them discusses more in general electromagnetic linear motion devices, which means also motorized actuators.

Therefore, it was necessary a detailed discussion about this topic. All the content of this book has been studied and developed by the author in his experience as designer of solenoid actuators and valves in aerospace industry. The computational techniques herein discussed have been developed by the author himself. Those techniques has given the author the possibility to successfully design and size solenoid devices.

The book is organized in the following chapters:

1. Sizing of the solenoid
2. Coils Winding

3. Materials and saturation
4. Complete calculation
5. Finite element method
6. Thermal analysis of the solenoid
7. Advanced calculations

In the appendixes:

- Appendix A: sensitivity analysis
- Appendix B: rectifier principles

The author hopes you enjoy the reading of the book.

Perugia (Italy), 9th April 2017

Daniele Righetti

Chapter 1

Sizing of the solenoid

1.1. Magnetic circuit theory

A solenoid actuator is basically a particular magnetic circuit. The definition of magnetic circuit comes from the Hopkinson Law, which is derived from the first Ampère Law:

$$\oint Hdl = \sum_{i}^{N} I_i \qquad (1.1.1)$$

where I_i are the currents flowing in the N coils, H is the magnetic field concatenated with the coils. The integration is made on any closed path crossing the coils. If we consider a solenoid made by N coils of the same wire, (1.1.1) becomes:

$$\oint Hdl = NI \qquad (1.1.2)$$

If we put these coils in air (or vacuum), we can observe the phenomena in Figure 1.1-1: we can observe that flux line of magnetic field are dispersed in the space around coils, while they are concentrated in the space near the axis.

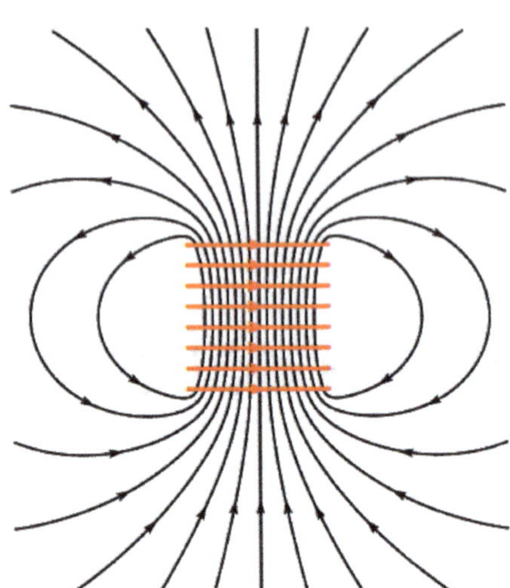

Figure 1.1-1: Solenoid field in air (vacuum).

If we wind the coils around a core of ferromagnetic material, which is a material with a very high magnetic permeability, the line flux we observe are very different from Figure 1.1-1 (see Figure 1.1-2a).

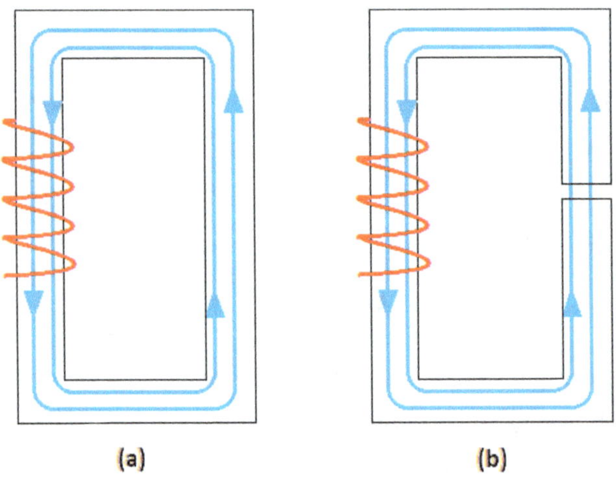

Figure 1.1-2: Coil winded on a ferromagnetic core.

This condition can be resumed in the following way: magnetic flux prefers to flow in ferromagnetic materials and this behaviour can be demonstrated with Maxwell equations. This characteristic gives the possibility to take control of magnetic field generated by current and to produce electromagnets. In fact if we now create an air gap in a point of the ferromagnetic material, we can observe that if the gap is little (with respect to the closed flux line length), the magnetic flux is approximately the same in the air gap and in the ferromagnetic material. The result is a uniform strong magnetic field across the gap and if we put for example a ferromagnetic steel across the gap, it will be attracted by one of the two sides of the gap: one side is the NORTH, the other one is the SOUTH. The computation of the magnetic field is described in the following.

Magnetic induction B and magnetic field H are not independent variables. In effect:

$$B = B(H) \tag{1.1.3}$$

For details regarding materials characteristics, refer to chapter 3. By now, we can consider a linear relation between B and H:

$$B = \mu H = \mu_r \mu_0 H \tag{1.1.4}$$

where μ is material magnetic permeability, μ_r is relative magnetic permeability, μ_0 is magnetic permeability of air and vacuum. The units of measure in the S.I. are resumed in Table 1.1-1:

Table 1.1-1: Ampère law relevant unit of measure

Symbol	Description	Unit of measure
B	Magnetic Induction	T (Tesla)
H	Magnetic Field	A/m (Ampère-Turns/meter)
μ_r	Relative Magnetic Permeability	-
μ_0	Magnetic Permeability of air (vacuum)	H/m (Henry/meter)
μ	Magnetic Permeability	H/m (Henry/meter)

Magnetic permeability of air is a constant:

$$\mu_0 = 4\pi \cdot 10^{-7}\,\frac{H}{m} \tag{1.1.5}$$

We can now write in a different way the (1.1.2):

$$\oint H dl = \oint \frac{B}{\mu}\,dl = \oint \frac{\Phi}{\mu S}\,dl = NI \tag{1.1.6}$$

where S is the area of the surface crossed by magnetic flux, $\Phi=BS$ is the flux. If we now consider that all the flux remains inside the device, Φ is a constant inside the operation of integration. If we now consider that along the path of integration there are different materials and different crossing surfaces we can write:

$$\Phi\left(\int_{1}\frac{dl_1}{\mu_1 S_1} + \int_{2}\frac{dl_2}{\mu_2 S_2} + ...\right) = NI \tag{1.1.7}$$

In (1.1.7) we call Reluctance the quantity:

$$\Re = \int \frac{dl}{\mu S} \tag{1.1.8}$$

The unit of measure of reluctance is H^{-1}. Now we can write:

$$\Phi(\Re_1 + \Re_2 + ...) = NI \tag{1.1.9}$$

and $\Re = \Re_1 + \Re_2.....$ the total reluctance and $f_m=NI$ the magneto-motive force. In this way, we can write the Hopkinson Law:

$$\Phi\Re = f_m \tag{1.1.10}$$

The fact that this formula is equivalent to Ohm law is the reason we call these devices magnetic circuits.

If we apply this concept to Figure 1.1-2(b), we observe that the magnetic circuit is composed by two different reluctances: the ferromagnetic material reluctance $\Re_m$, the air gap reluctance $\Re_a$:

$$\Re_m = \frac{l_m}{\mu_r \mu_0 S} \;;\; \Re_a = \frac{l_0}{\mu_0 S} \tag{1.1.11}$$

where l_m is the length of the path along ferromagnetic material and l_0 the length of the air gap; the crossing surface area is considered the same for both reluctances. Now we can write:

$$\Phi \frac{1}{\mu_0 S}\left(\frac{l_m}{\mu_r}+l_0\right)=NI \qquad (1.1.12)$$

Therefore, the magnetic flux is:

$$\Phi=\frac{\mu_0 S \cdot NI}{\left(\dfrac{l_m}{\mu_r}+l_0\right)} \qquad (1.1.13)$$

The magnetic induction is:

$$B=\frac{\mu_0 NI}{\left(\dfrac{l_m}{\mu_r}+l_0\right)} \qquad (1.1.14)$$

From (1.1.14) we can observe that with a good ferromagnetic material (high relative permeability) and a low air gap we can obtain very high value for magnetic induction. The formula (1.1.14) is often simplified considering that relative permeability of ferromagnetic materials can be of the order of $\mu_r = 400 \div 1000$ (pure iron can be 8000) and so we can write:

$$B=\frac{\mu_0 NI}{l_0} \qquad (1.1.15)$$

It is important to underline that this formula is not valid when the air gap has a more complicated geometry, for example for conical plunger solenoid (see chapter 5). Therefore, (1.1.14) and (1.1.15) are valid only for plane keeper solenoid.

1.2. Sizing of a solenoid

The sizing of a solenoid is based upon the choice of input parameters and upon the choice of a geometrical model, which represents real solenoid. In the present book, we will describe the sizing of a solenoid represented by Figure 1.2-1.

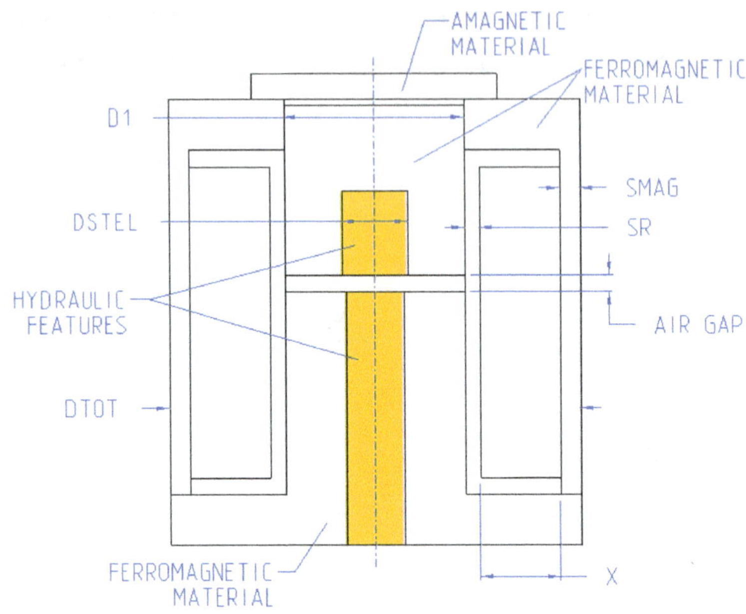

Figure 1.2-1: Solenoid geometry.

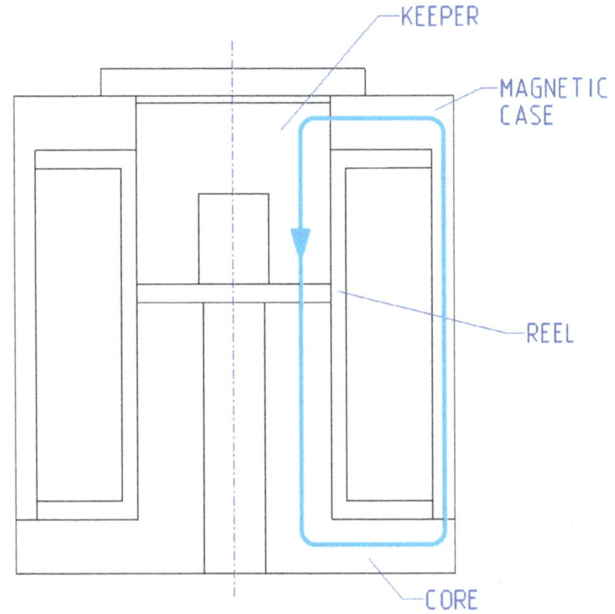

Figure 1.2-2: Solenoid main components and magnetic circuit representation.

The solenoid in Figure 1.2-1 is a typical pilot valve. In this representation, hydraulic features are not represented because at the moment we are interested to electromagnetic sizing. The main output is the magnetic force whose target value depends form the configuration of hydraulic features. The valve can be normally opened or normally closed and so on. It is up to the designer fixing target values. Moreover, this general representation gives the possibility to size all devices having plane face keeper, for example electro-valves or short stroke solenoid actuators. Solenoid with keeper with different geometry cannot be sized with this procedure.

The input parameters are the following:

- Temperature range: minimum temperature T_{MIN}, maximum temperature T_{MAX}, nominal temperature T_0. Nominal temperature is usually equal to 20°C. If a different choice is made, the value of temperature coefficient and resistivity of wire material has to be changed accordingly. This operation is not suggested because all standard magnet wire tables refers to values at 20°C. Refer to §2.7 for details. For this reason, nominal temperature is considered as a reference temperature.

- Power supply range: minimum voltage V_{MIN}, maximum voltage V_{MAX}, nominal voltage V_{NOM}. This choice means that in real operation the power control is based upon voltage. In some cases, current is controlled and a different formulation of the problem is needed.

- Wire characteristics: solenoid coils are made of magnet wire. This kind of wire is basically a copper wire enamelled with a polymeric insulant that avoids contact between conducting part of coils, giving the possibility of winding coils in a very compact size. The information needed is the conductor diameter D_{IW}, the external diameter D_W, linear resistance of wire R_{LIN} (at the nominal temperature). Refer to [1] for details about wire characteristics.

- Geometrical characteristics: keeper diameter D_1, thickness of reel S_R and magnetic case S_{MAG}, number of coils in radial direction N_S and length of the air gap l_0. Another important geometrical parameter is the winding efficiency ε.

- Maximum current density ρ_{MAX}: this is the input parameter that define thermal performances of the solenoid. If this parameter has a high value, it means that we can reduce coil length and so solenoid dimension but, in the same time, we have more power to dissipate by Joule effect.

- Insulation parameters: an insulant fabric is placed between reel and coils to prevent contact between wires and reel. Its thickness S_{ISOL} shall be considered in the computation of solenoid size. Another insulant material is placed between coils and magnetic case. The thickness S_{ADJ} of this insulation shall be modelled.

The sizing of the solenoid is made with the following steps:

1. From the value assumed for ρ_{MAX} and from the choice of the wire we can calculate the nominal current I_{NOM} with the following formula:

$$I_{NOM} = \rho_{MAX} \frac{\pi}{4} D_{IW}^2 \tag{1.2.1}$$

2. From the value of **V_NOM** using the Ohm law we can calculate nominal resistance **R₀**:

$$R_0 = \frac{V_{NOM}}{I_{NOM}} = \frac{4V_{NOM}}{\rho_{MAX}\,\pi D_{IW}^2} \tag{1.2.2}$$

3. From the choice of the wire, we know the linear resistance **R_LIN**. So we can compute the length of wire by:

$$L = \frac{R_0}{R_{LIN}} \tag{1.2.3}$$

4. In §2.3 the concepts of winding efficiency and coils positioning angle are explained. We can write:

$$\alpha = 90 - 30\varepsilon\,[\text{deg}] \tag{1.2.4}$$

5. If the number of windings in the radial direction is N_S, we can calculate the radial size X of coils (refer to §2.4 for details):

$$X = D_w + (N_S - 1)D_w \sin(\alpha) \tag{1.2.5}$$

The height of the winding zone is (refer to §2.4 for details):

$$Y = \frac{LD_W}{\pi N_S (D_1 + 2S_R + 2S_{ISOL} + X)} \tag{1.2.6}$$

In the inner diameter of the coils, an insulant fabric of thickness equal to S_ISOL is installed. In the outer diameter, an insulation layer is required: it is composed by a sealant and in some case even by the same fabric insulant installed in the inner diameter (refer to §2.2 for details). Sealant is required in the case of solenoid valve to protect coils from hydraulic fluid contamination. The thickness of this layer is represented by the parameter S_ADJ. Therefore, we can define a total radial size of winding zone:

$$X_{TOT} = X + S_{ISOL} + S_{ADG} \tag{1.2.7}$$

With these formulas, we can define the inner diameter of the coil **D$_{INN}$** and the outer diameter **D$_{OUT}$**:

$$D_{INN} = D_1 + 2S_R \tag{1.2.8}$$

$$D_{OUT} = D_1 + 2S_R + 2X_{TOT} \tag{1.2.9}$$

On the top and on the bottom of the winding zone is installed an insulation fabric. In this way, the coils are insulated from the reel. Moreover we have to consider that at the beginning of the winding process the first coil is protected with a heat shrinkable tubing, because this coil is one of the lead of the entire winding (refer to §2.2 for details). Therefore, we can define a total height of the winding zone:

$$Y_{TOT} = Y + 2S_{ISOL} + D_W + 2W_{TH} \tag{1.2.10}$$

where **W$_{TH}$** is the radial thickness of the heat shrinkable tubing.

6. Now the computation of total number of coils is possible. In effect from (1.2.6) we can write the number of windings in the **Y** direction **N$_Y$**:

$$N_y = \frac{Y}{D_W} \tag{1.2.11}$$

Therefore, we can write that the total number **N** is:

$$N = N_S N_Y \tag{1.2.12}$$

7. The computation of the force is based upon magnetic circuit consideration explained before. From the definition of the Maxwell Stress Tensor (see §1.4) we can write the magnetic force acting on the keeper as:

$$F = \frac{B^2 S}{2\mu_0} \tag{1.2.13}$$

where S is the area of the section of the keeper crossed by magnetic flux. Looking at the Figure 1.2-1 this area is equal to:

$$S = \frac{\pi\left(D_1^{\,2} - D_{STEL}\right)}{4}$$ (1.2.14)

All the crossing area of the magnetic case represented in Figure 1.2-2 must be equal or superior to this value. Therefore, the choice of the parameter **S$_{MAG}$** is constrained by the fact that magnetic case crossing area **A$_C$** shall be:

$$A_C = \frac{\pi\left(D_{TOT}^2 - D_{ESTR}^2\right)}{4} \geq S$$ (1.2.15)

Where D$_{TOT}$ is the total external diameter of the solenoid:

$$D_{TOT} = D_{ESTR} + 2S_{MAG}$$ (1.2.16)

So we can write:

$$F = \frac{\mu_0\left(NI\right)^2 S}{2\left(\dfrac{l_m}{\mu_r} + l_0\right)^2}$$ (1.2.17)

8. In the computation of the magnetic force with (1.2.17) the induction field is calculated with (1.1.14), where the length of the path along ferromagnetic material *l$_m$*, referring to Figure 1.2-1, can be approximated as:

$$l_m = 2X + 2Y - l_0 \approx 2X + 2Y$$ (1.2.18)

Looking at the formula (1.2.17), we can see that magnetic force depends from current and not from voltage. In this formulation, we consider a control in terms of voltage. Therefore, current depends from input voltage and temperature, because resistivity depends from temperature:

$$R(T) = R_0\left(1 + \alpha\left(T - T_0\right)\right)$$ (1.2.19)

where T_0 is the reference temperature for R_{LIN} in (18) (typically 20°C) and α is the coefficient of temperature of wire material at T_0 (refer to §2.7 for details).

1.3. Sensitivity analysis

It is very important to understand deeply the influence of the input parameter on the solenoid performance. When a sizing is required, typically, a target force is defined and the solenoid designer shall be able to find the correct sizing in order to "produce" that target force and to respect constrains in terms of dimensions and weight. Understanding the most sensible parameters is very important in order to adjust the design.

In order to perform a sensitivity analysis is necessary to write the expression of the force in (1.2.17) in such a way that all input parameters appears explicitly. From (1.2.17), we will start from the variable l_m expressed in (1.2.18) and focus on the expression of Y expressed in (1.2.6). We obtain an explicit expression to the variable L substituting (1.2.2) in (1.2.3):

$$L = \frac{4V_{NOM}}{R_{LIN}\rho_{MAX}\pi D_{IW}^2} \tag{1.3.1}$$

Substituting (1.2.4) in (1.2.5) (writing the angle α in radians) and (1.2.5) and (1.3.1) in (1.2.6) we can write:

$$Y = \frac{4V_{NOM}D_W}{R_{LIN}\rho_{MAX}D_{IW}^2\pi^2 N_S\left(D_1 + 2S_R + 2S_{ISOL} + D_W + (N_S-1)D_W\sin\left((90-3\varepsilon)\frac{\pi}{180}\right)\right)} \tag{1.3.2}$$

Now we can write the explicit expression for N:

$$N = N_s N_y = N_s\frac{Y}{D_W} \tag{1.3.3}$$

$$N = \frac{4V_{NOM}}{R_{LIN}\rho_{MAX}D_{IW}^2\pi^2\left(D_1 + 2S_R + 2S_{ISOL} + D_W + (N_S-1)D_W\sin\left((90-3\varepsilon)\frac{\pi}{180}\right)\right)}$$

Now we express the current absorbed in terms of actual voltage V and temperature T using (1.2.2):

$$I = \frac{V}{R} = \frac{V}{R_0\left(1 + \alpha(T - T_0)\right)} = \frac{V \rho_{MAX} \pi D_{IW}^2}{4 V_{NOM}\left(1 + \alpha(T - T_0)\right)} \qquad (1.3.4)$$

The product NI:

$$NI = \frac{V}{R_{LIN} \pi^2 \left(D_1 + 2S_R + 2S_{ISOL} + D_W + (N_S - 1)D_W \sin\left((90 - 3\varepsilon)\frac{\pi}{180}\right)\right)\left(1 + \alpha(T - T_0)\right)}$$

$$(1.3.5)$$

Substituting (1.2.5) and (1.3.2) in (1.2.18) and then substituting (1.2.18), (1.3.5) and (1.2.14) in (1.2.17) we obtain the explicit formula for magnetic force:

$$F = \cfrac{\cfrac{\mu_0}{8}\left(D_1^2 - D_{STEL}^2\right)\cfrac{V^2}{R_{LIN}^2 \pi\left(D_1 + 2S_R + 2S_{ISOL} + D_W + (N_S - 1)D_W \sin\left((90 - 3\varepsilon)\frac{\pi}{180}\right)\right)^2\left(1 + \alpha(T - T_0)\right)^2}}{\left(\left(l_0 + \cfrac{1}{\mu_r}\left(2\cfrac{D_W + (N_S - 1)D_W \sin\left((90 - 3\varepsilon)\frac{\pi}{180}\right) + 4V_{NOM} D_W}{R_{LIN}\rho_{MAX} D_{IW}^2 \pi^2 N_S\left(D_1 + 2S_R + 2S_{ISOL} + D_W + (N_S - 1)D_W \sin\left((90 - 3\varepsilon)\frac{\pi}{180}\right)\right)} - l_0\right)\right)\right)^2}$$

$$(1.3.6)$$

A detailed analysis is shown in appendix A for a typical electro-valve solenoid. The main result is that there are two parameters that have a higher sensitivity than others. The first one, in order of importance, is the wire size: increasing the wire size (following tables of international specifications) increases magnetic force by a value of 50% approximately. The second parameter is the keeper diameter: increasing the diameter by a value of 1 mm produces an increase in the magnetic force by a value of 10%.

Resuming the results: increasing wire size gives the possibility to obtain a solenoid of superior performances from the starting design; increasing keeper diameter gives the possibility to adjust the design.

1.4. Maxwell Stress Tensor

From Maxwell equations, it is possible to express a stress tensor describing the mechanical stress acting on a generic surface positioned in an electromagnetic field. If the field is only magnetic, we can write the expression of a generic element of the tensor as:

$$\sigma_{ij} = \frac{1}{\mu_0} B_i B_j - \frac{1}{2\mu_0} B^2 \delta_{ij} \tag{1.4.1}$$

where is δ_{ij} the Kronecker delta. The expression is valid only in air or vacuum. We use (1.4.1) considering that:

- We choose as surface for computing the stress the bottom side of keeper;
- We evaluate the expression in the air gap where it is valid;
- Thanks to the continuity values condition for all field variables, the expression represents the stress acting on the bottom side of keeper.

If we call "x" the solenoid axis, if we use the properties of Kronecker delta and the fact that field B has only the "x" component, we write:

$$\sigma_{xx} = \frac{1}{\mu_0} B_x B_x - \frac{1}{2\mu_0} B^2 \delta_{xx} = \frac{1}{\mu_0} B^2 - \frac{1}{2\mu_0} B^2 = \frac{B^2}{2\mu_0} \tag{1.4.2}$$

If we consider that, this stress is uniform on the bottom surface of keeper we can write:

$$F = \sigma S = \frac{B^2 S}{2\mu_0} \tag{1.4.3}$$

This expression is general; it is valid even in saturated conditions.

Chapter 2

Coils Windings

2.1. Magnet wire

The wires used for solenoid coils are called magnet wire. These wires are used in all devices that provide bobbins and winding procedure, such as solenoid, electrical motors and transformers. This is because their insulation is given by a very thin layer of insulant material. For a wire with a diameter of the conductor of 0.18 mm, this layer has a thickness typically of 0.02 mm. This key feature gives the possibility to create bobbins with a very reduced size.

There are many international standards that define performances and dimensions for magnet wire. In this book, we mention the IEC 60317-13 (see [1]), or the equivalent NEMA MW 35-C (see [2]). This wire has the following features:

- It is a circular type;
- Conductor material: copper;
- Insulation: is a (THEIC) polyesterimide overcoated with polyamide-imide;
- Temperature Index: 200. It means that wire can work for an indefinite time at 210 °C and it can resist for 5000 hours at

230°C. For very short time it can work at 340 °C which is its thermoplastic temperature above which polymeric insulation is melted;

- Three different grade: basically, they differ on the thickness of the insulant.

In the tables of the IEC 60317-13 relevant properties are:

- External diameter variation;
- Linear resistance variation;
- Maximum tension of wire during winding process.

2.2. Winding procedure

Winding procedure can be realized with different kind of machines, manual, automatic or semiautomatic machines, it depends from industrialization choices. A typical semiautomatic machine is represented in Figure 2.2-1:

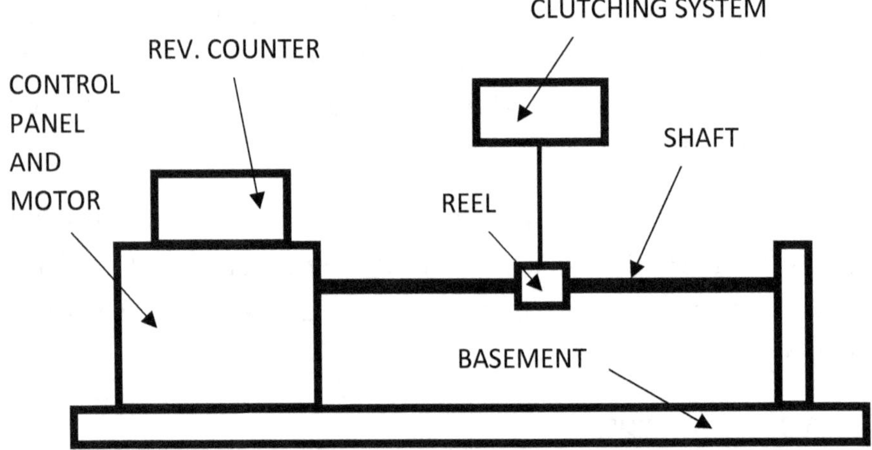

Figure 2.2-1: A typical semiautomatic winding machine.

This kind of machine has two main feature: a rev. counter and a clutching system. The rev. counter has the function to count the number of coils to wind: this parameter shall be prescribed in the drawing of reel assembly. The clutching system has the important function to control the tension of wire during winding. The maximum tension is a requirement of the international standards and it is very important to respect: too high tension during winding process can cause a stretching of wire and so the decrease of wire diameter. The result is a wire with higher resistance, a lower current absorption and so lower magnetic force. Another important consequence of a high tension of the wire will be discussed in §2.6.

Now it is easy to understand that controlling number of coils and maximum wire tension ensure to respect prescribed parameter N and I: the quantity (NI) is correct and the solenoid will be able to "produce" the design magnetic force.

It follows a very short description of winding process.

1. Insulant fabric assembly: an insulant adhesive fabric, typically a PTFE impregnated fiberglass, is assembled in the positions indicated by Figure 2.2-2. We mention the fabric **LUBRIGLAS®** by **ANGST+PFISTER®** or **KAPTON®** by **DuPont®**. It is important to underline that the first fabric to be assembled is the inner one: this choice ensures that the points shown in Figure 2.2-2 cannot get in contact with coils. In this way reel is completely insulated from winding.

2. Positioning of the first lead: first lead is positioned as shown in Figure 2.2-3. The lead must get in the winding zone from a typical slot designed on the reel. The use of a heat shrinkable tubing is required to protect the wire because, as it can easily understand, all coils will be positioned above it. This operation is very important to obtain a good and reliable winding.

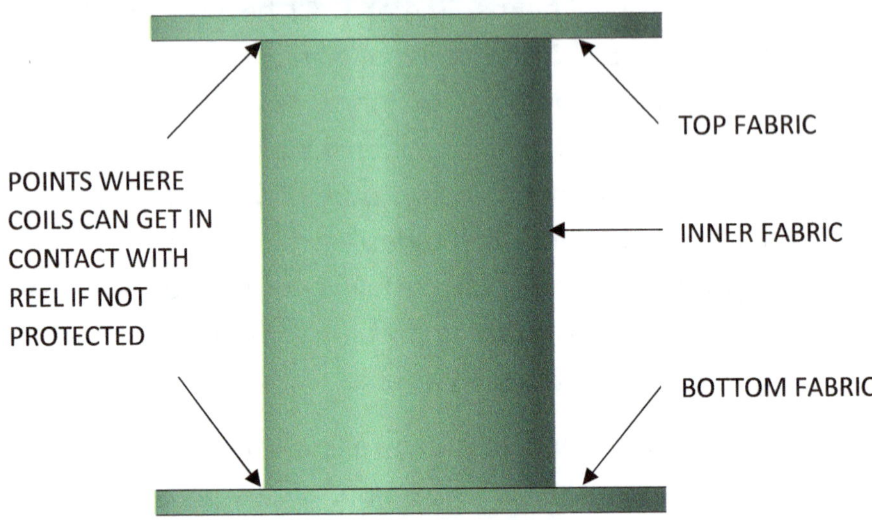

Figure 2.2-2: Winding procedure: insulant fabric application.

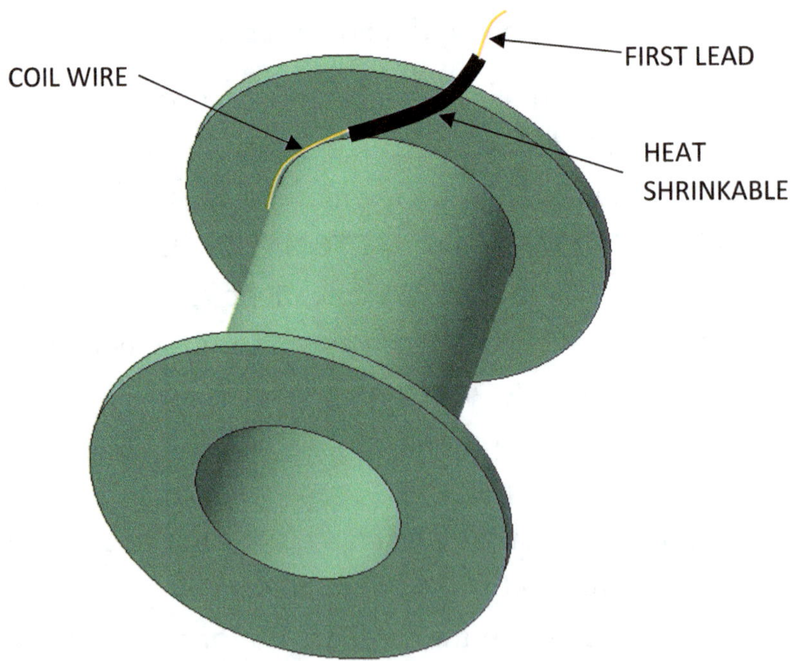

Figure 2.2-3: Winding procedure: first lead protection.

3. The coils are winded around the reel paying attention to final number of coils prescribed by drawing and wire tension. The final position of coils will be discussed in §2.3.

4. When the final number of coils has been winded, the other lead of the wire is positioned in the same slot used for the first lead and it is protected with a heat shrinkable tubing. After this operation, the two leads are tin welded with stranded wires (typically American Wire Gauge type). Welding zone shall be protected with heat shrinkable tubing to provide stiffness and to avoid the loss of the joint. The choice of size of stranded wires shall be in accordance with maximum current rating prescribed by wire specification.

5. The stranded wires shall be fastened (using fastening wire and/or adhesive). This operation is very important because during solenoid assembly the two terminals can be stretched and/or twisted: a good fastening system avoids magnet wire damage.

   If an electrical connector is used the operation of fastening is not strictly necessary. Let us say that it is a recommended practice in solenoid production for reliability and safety reasons.

6. The assembly of reel and coils is installed inside the magnetic case (see Figure 1.2-2). In this operation is a recommended practice to protect windings with insulant fabric even if the insulation between coils and magnetic case is made by the use of sealant. That is because during this operation there is always the risk to damage wire. The sealant used is typically a silicon one because it can work at temperature compatible with wire

thermal class of 200. We mention the sealant **PR1930-2** of **Le Joint François**®, suitable for high temperatures.

2.3. Winding efficiency

In winding process, the disposition of coils can be of various type. Figure 2.3-1 shows that there can be two opposite disposition: "perfect disposition", characterized by a positioning angle α of 60 degrees and the so-called "square disposition", characterized by a positioning angle of 90 degrees. The first one is the disposition that best uses the winding area, vice versa the second one. Figure 2.3-1b shows and intermediate disposition with a positioning angle of 70 degrees.

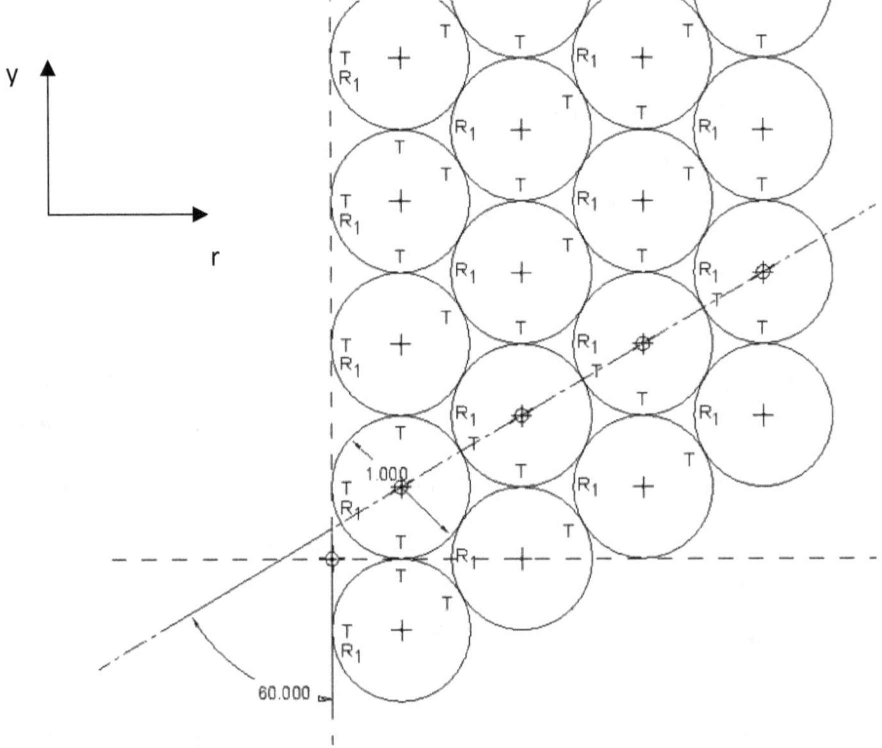

(a) Perfect disposition.

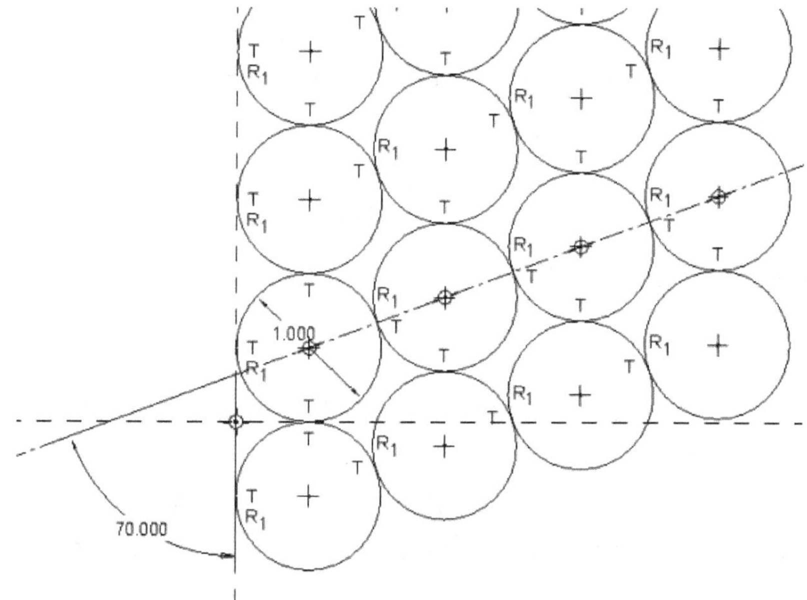

(b) Intermediate disposition.

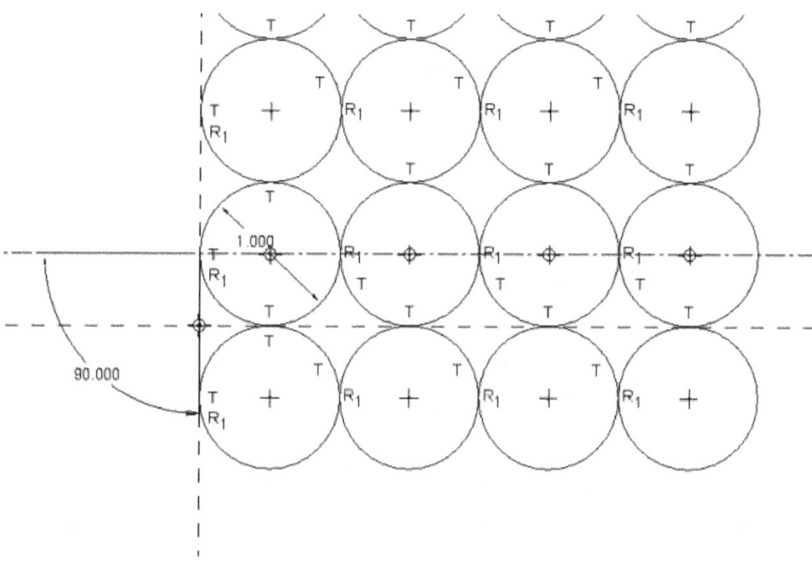

(c) Square disposition.

Figure 2.3-1: Winding disposition.

So we can define a winding coefficient or winding efficiency ε, in such a way that if α=60 deg, the efficiency will be ε=1; if α=90 deg, the efficiency will be ε=0. So we can write:

$$\alpha = (90 - 30\varepsilon)\frac{\pi}{180}\left[rad\right] \qquad (2.3.1)$$

We define the "Filling Factor" F:

$$F = A_W / A_T \qquad (2.3.2)$$

where A_W is the coils area and A_T is the total area. The total area is a rectangle of sides X and Y: $A_T = X*Y$.

From Figure 2.3-1, it is easy to understand that, if N_y is the number of windings in the longitudinal direction and N_s in the radial direction, the side Y is:

$$Y = N_y D_W \qquad (2.3.3)$$

where D_W is the wire external diameter. The side X (refer to §2.4 for details):

$$X = D_W + (N_S - 1)D_W \sin(\alpha) = D_W\left[1 + (N_S - 1)\sin(\alpha)\right] \qquad (2.3.4)$$

The coils area is:

$$A_W = N_S N_y \frac{\pi D_W^2}{4} \qquad (2.3.5)$$

We can write the "Filling factor" F:

$$F = \frac{N_S N_y \dfrac{\pi D_w^2}{4}}{N_y D_w^2\left[1 + (N_S - 1)\sin(\alpha)\right]} = \frac{N_S \pi}{4\left[1 - \sin(\alpha) + N_S \sin(\alpha)\right]} \qquad (2.3.6)$$

We are interested to understand the link between ε and F, so we want to know an average value for F: we consider an infinite winding area and an infinite number of coils. In this way, we eliminate boundary effect of rectangle sides on "Filling factor" and we obtain an expression that depends only on the kind of winding disposition:

$$F = \lim_{N_S \to \infty} \frac{N_S \pi}{4[1 - \sin(\alpha) + N_S \sin(\alpha)]} \tag{2.3.7}$$

which is equal to:

$$F = \frac{\pi}{4\sin(\alpha)} \tag{2.3.8}$$

The formula (2.3.8) is very important because gives a simple interpretation of winding efficiency.

The following graph plots the function described by (2.3.8).

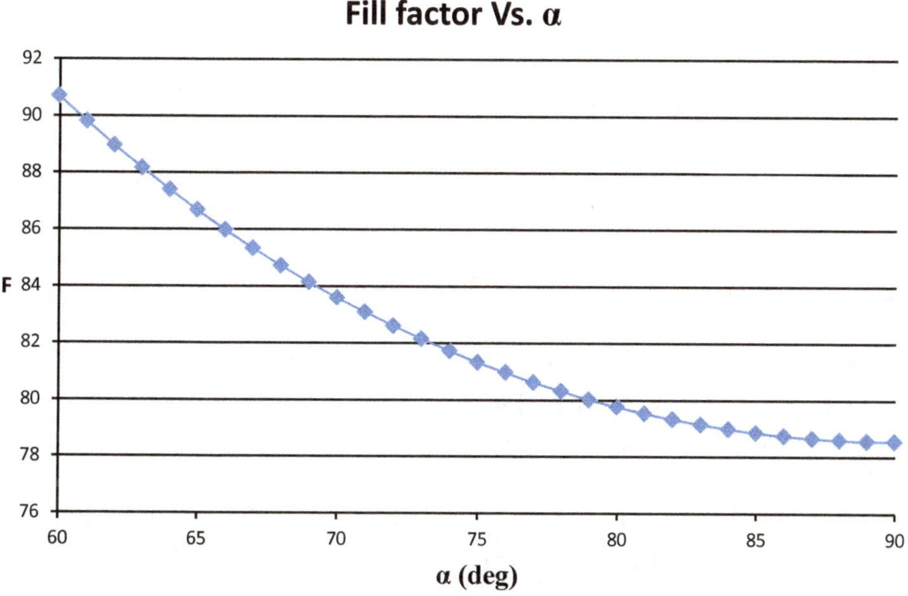

Figure 2.3-2: F Vs α.

In the following graph, filling factor is plotted against efficiency ε using the dependency described by equation (2.3.1) and substituting in the (2.3.8).

It is easy to understand that a "perfect disposition" with ε=1 guarantees an higher filling factor.

Fill factor Vs. ε

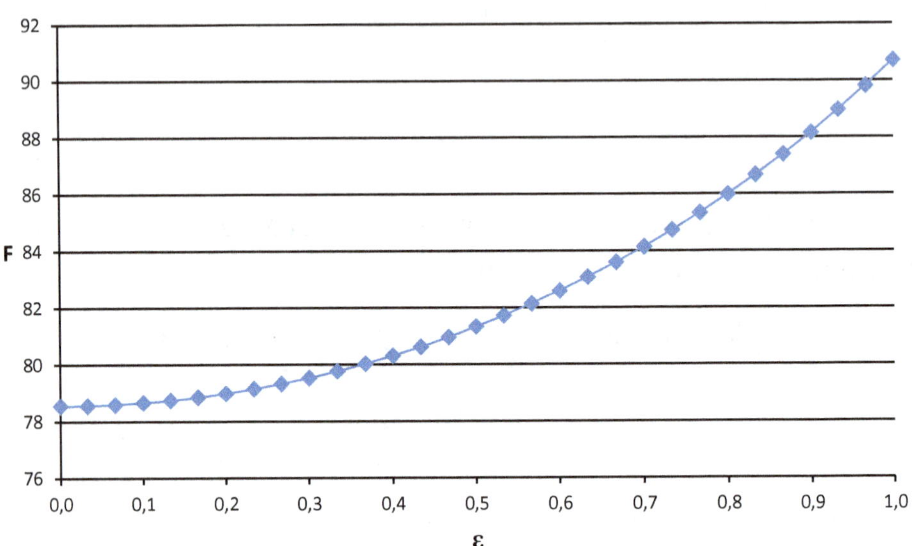

Figure 2.3-3: F Vs ε.

2.4. Computation of winding size

In chapter 1, we have described the sizing procedure of a solenoid. We have used the formulas (1.2.5) and (1.2.6) to compute respectively the radial size and height of winding zone.

In Figure 2.4-1, the scheme used to find (1.2.5) is shown. In this case, we can write the following relations:

$$N_S = 4$$

$$X = \frac{D_W}{2} + 3x_3 + \frac{D_W}{2} = D_W + (N_S - 1)x_3 \qquad (2.4.1)$$

$$x_3 = D_W \sin(\alpha)$$

So we can write:

$$X = D_W + (N_S - 1)D_W \sin(\alpha) \qquad (1.2.5)$$

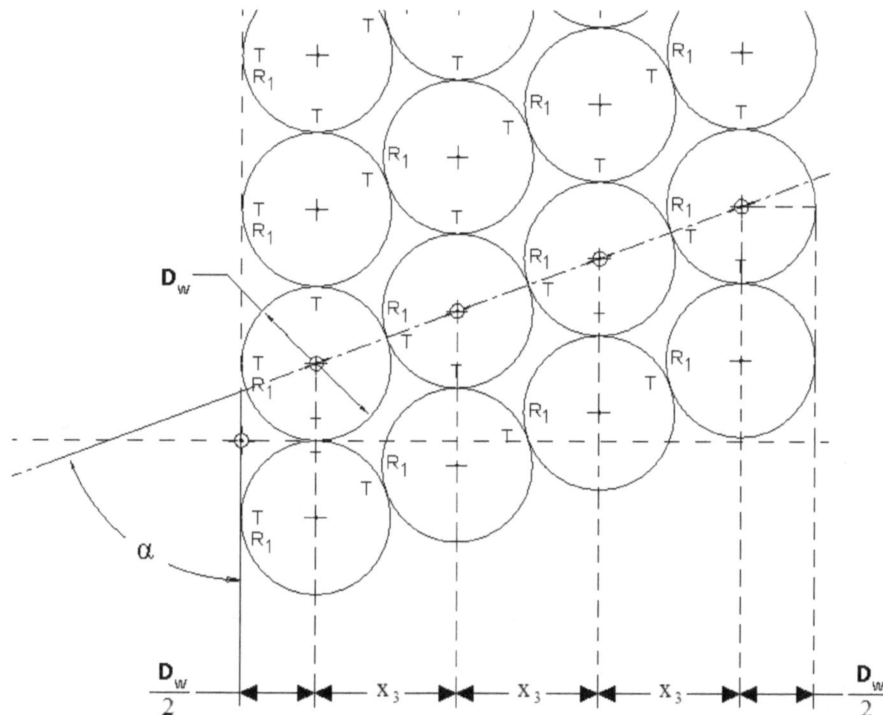

Figure 2.4-1: Scheme for the computation of X.

To obtain the formula (1.2.6) we have to recall the second Guldino theorem for the volume of solids of revolution:

$$V = 2\pi A d \qquad (2.4.2)$$

where A is the area of the surface we want to rotate and d is the distance of its geometric centroid from the axis of revolution. In this case, the area is a rectangle of sides X and Y as shown in the Figure 2.4-2.

We can write that:

$$d = \frac{D_2 + X}{2} = \frac{D_1 + 2S_R + 2S_{ISOL} + X}{2} \qquad (2.4.3)$$

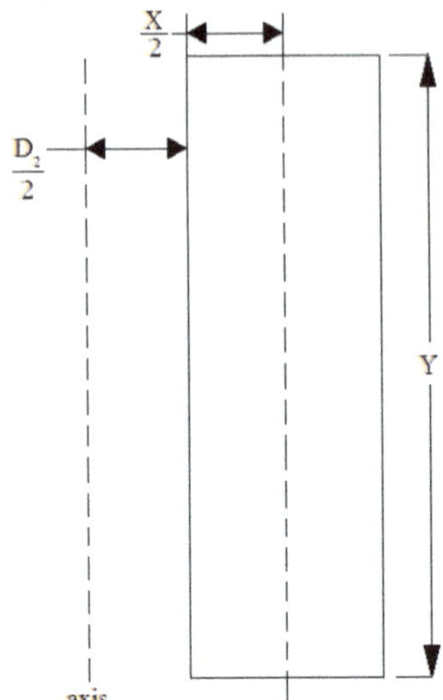

Figure 2.4-2: Scheme for the computation of Y.

The area A is composed by all coils:

$$A = N_S N_Y A_f \qquad (2.4.4)$$

where A_f is the section area of the wire. Considering that $N_y = Y/D_w$, we can write:

$$V = 2\pi N_S N_Y A_f \frac{D_1 + 2S_R + 2S_{ISOL} + X}{2} = 2\pi N_S \frac{Y}{D_W} A_f \frac{D_1 + 2S_R + 2S_{ISOL} + X}{2}$$

$$(2.4.5)$$

We can write the volume as:

$$V = LA \qquad (2.4.6)$$

If we equal the (2.4.5) and (2.4.6) and explicit the height H we obtain the (1.2.6):

$$Y = \frac{LD_W}{\pi N_S \left(D_1 + 2S_R + 2S_{ISOL} + X \right)} \qquad (1.2.6)$$

2.5. Number of coils Vs. Wire diameter

In chapter 1, we used the wire external diameter D_W to size the solenoid and to find the total number of coils to be winded. It is easy to understand that the value of this diameter has its own tolerance prescribed by international standards. Therefore, D_W represents the maximum wire external diameter. We call D_{WINT} the minimum value of D_W taken from tables. This difference is due to variation of the thickness of enamelled insulant layer.

It is easy to understand that if two wire of the same length L and diameter respectively of D_W and D_{WINT} are winded around the same reel, we will count a different number of coils. In particular, we will notice that for the wire of D_{WINT} we will count a greater number of coils; it means that we will have a greater force for the solenoid because N is greater while the absorbed current is the same because the length and so the resistance are the same.

The operator does not control the length of the wire but controls the number of windings. Therefore, it is necessary to give the operator the number of coil, depending from wire diameter, which guarantees the correct length of the wire and so the correct absorbed current.

Now we have to find the relation between N and wire diameter. Input data are:

- L: wire length;
- Y: winding height (computed with maximum D_W: see solenoid sizing in chapter 1);
- $D_2 = D_1 + 2S_R + 2S_{ISOL}$: internal diameter of windings zone;
- α: positioning angle of coils.
- D_W: wire diameter (a value between D_W and D_{WINT}).

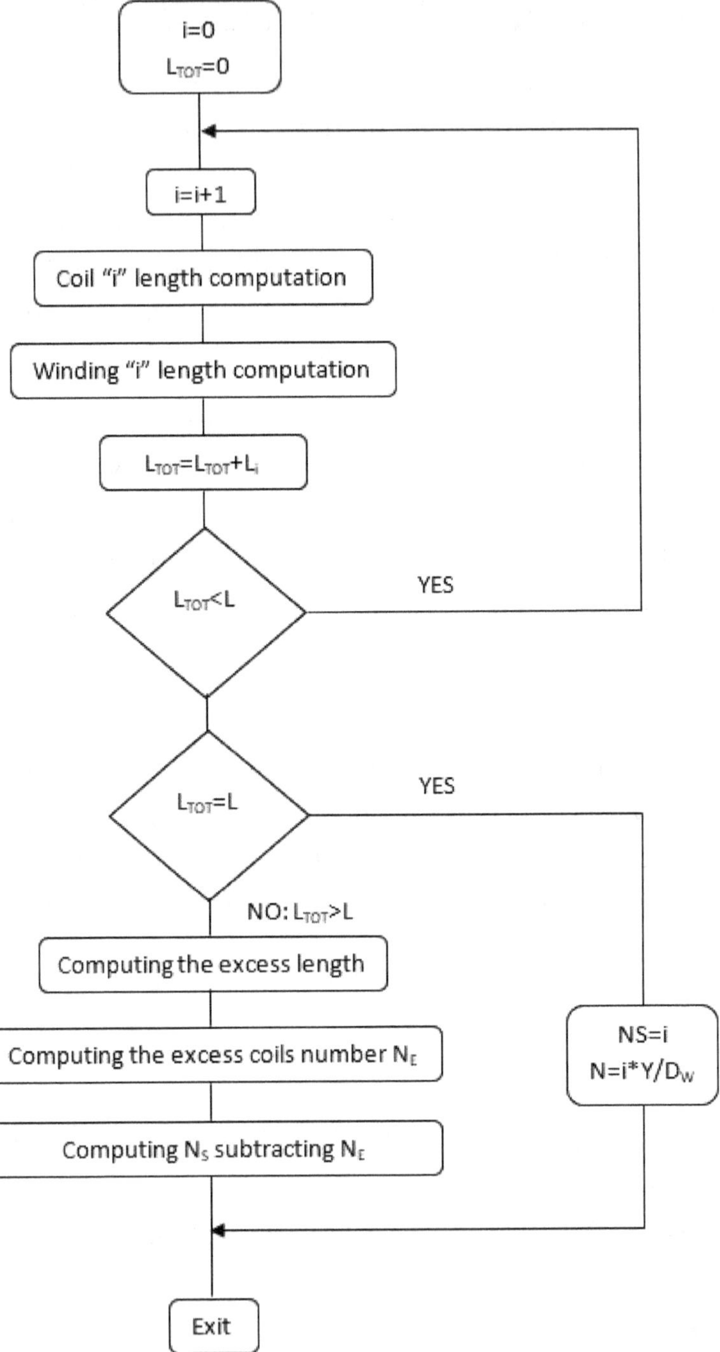

Figure 2.5-1: N Vs. D_W block diagram.

An iterative procedure is applied with a logical scheme described by the block diagram shown in Figure 2.5-1. The procedure consists in counting the number of coils that provide the wire length L, which is computed with helix curve formulas (the helix pitch is D_w). Let us imagine winding a first layer from the bottom to the top of the reel: we know from helix formulas the length of the layer; after the wire has been winded, we have to increase the diameter of the helix because the second layer is winded around the first layer and so on.

With this procedure is possible to give the operator a diagram of number of coils Vs. wire diameter (see Figure 2.5-2): the operator measure the wire and with the diagram chooses the correspondent number of coils. One can demonstrate that the diagram is linear.

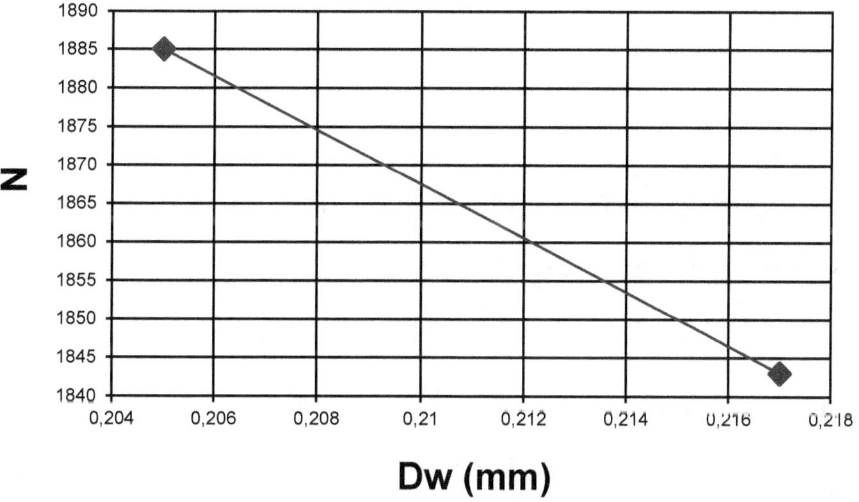

Figure 2.5-2: N Vs. D_w diagram.

2.6. Reel deformation during winding

Another effect due to an incorrect control of wire tension is the deformation of reel. During winding process reel is compressed by wire tension. This effect is usually negligible but, in the case that reel wall is too thin, it is possible that reel is deformed. This condition must be avoided because it gives rise to incorrect keeper movement and so to incorrect solenoid operation.

To better understand this phenomenon, we have to build a mathematical model. Figure 2.6-1 shows how tension, in an infinitesimal part of wire, produces a pressure on the reel surface.

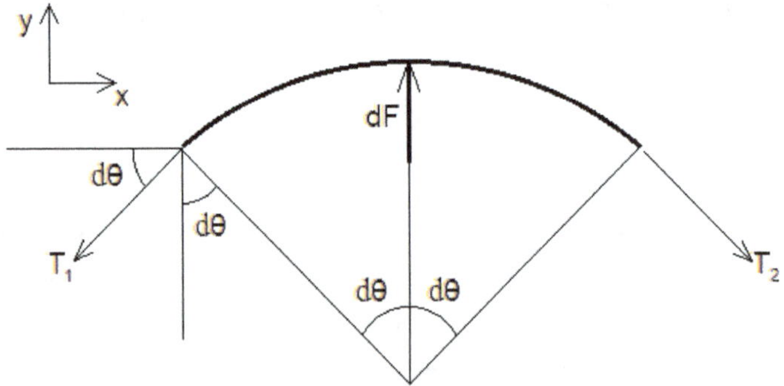

Figure 2.6-1: Free body of an infinitesimal part of wire.

If we consider the force equilibrium on x) direction:

$$x): T_1 \cos(d\vartheta) = T_2 \cos(d\vartheta) \rightarrow T_1 = T_2 = T \tag{2.6.1}$$

Using the above result, we write the y) force equilibrium:

$$y): 2T \sin(d\vartheta) = dF \rightarrow dF = 2T d\vartheta \tag{2.6.2}$$

If we express the infinitesimal force with pressure, we find the following formulas:

$$dF = \tilde{p} r d\vartheta = 2T d\vartheta \tag{2.6.3}$$

$$\tilde{p} = \frac{2T}{r} \qquad (2.6.4)$$

The parameter " $\tilde{p}$ " in (2.6.4) has the dimension of a pressure per unit length. The pressure acting on the reel for a layer of coils is:

$$p_S = \tilde{p}\,\frac{N_y}{Y} \qquad (2.6.5)$$

Considering that $N_y = Y/D_W$:

$$p_S = \frac{2T}{r}\,\frac{Y}{D_W}\,\frac{1}{Y} = \frac{2T}{rD_W} \qquad (2.6.6)$$

The formula (2.6.6) represents the pressure of a layer acting on cylinder with radius "r". For the first layer "r" is equal to $D_2/2$ while for the other layers this radius is the radius of the previous layer. Therefore, we can write the total pressure as:

$$p = \frac{2T}{D_W}\sum_{i=0}^{N_S-1}\frac{1}{r_i} \qquad (2.6.7)$$

The radius $r_0 = D_2/2$, while r_i:

$$r_i = r_0 + \frac{D_W}{2}\left[1 + (2i-1)\sin\alpha\right], \quad 1 \le i \le N_S - 1 \qquad (2.6.8)$$

So we can write:

$$p = \frac{2T}{D_W}\left(\frac{1}{r_0} + \sum_{i=1}^{N_S-1}\frac{1}{r_i}\right) = \frac{2T}{D_W r_0}\left(1 + \sum_{i=1}^{N_S-1}\frac{r_0}{r_i}\right) \qquad (2.6.9)$$

The summation is:

$$\sum_{i=1}^{N_S-1}\frac{r_0}{r_i} = \sum_{i=1}^{N_S-1}\frac{r_0}{r_0 + \dfrac{D_W}{2}\left[1+(2i-1)\sin\alpha\right]} = \sum_{i=1}^{N_S-1}\frac{1}{1 + \dfrac{D_W}{D_2}\left[1+(2i-1)\sin\alpha\right]}$$

$$(2.6.10)$$

Therefore, the total pressure acting on the surface of the reel is:

$$p = \frac{4T}{D_W D_2}\left(1 + \sum_{i=1}^{N_S-1} \frac{1}{1 + \dfrac{D_W}{D_2}\left[1 + (2i-1)\sin\alpha\right]}\right) \qquad (2.6.11)$$

If we consider a wire per IEC 60317-13 with conductor diameter of 0.18 mm from tables we have:

$$D_W = 0.217mm$$

$$T_{MAX} = 0.225kg = 2.207N$$

For a solenoid with N_S=23, using (24) we have:

$$p = 54.8\ MPa$$

If we apply this uniform pressure on the reel surface and we solve the static problem with a FEM software, we obtain the results shown in Figure 2.6-2.

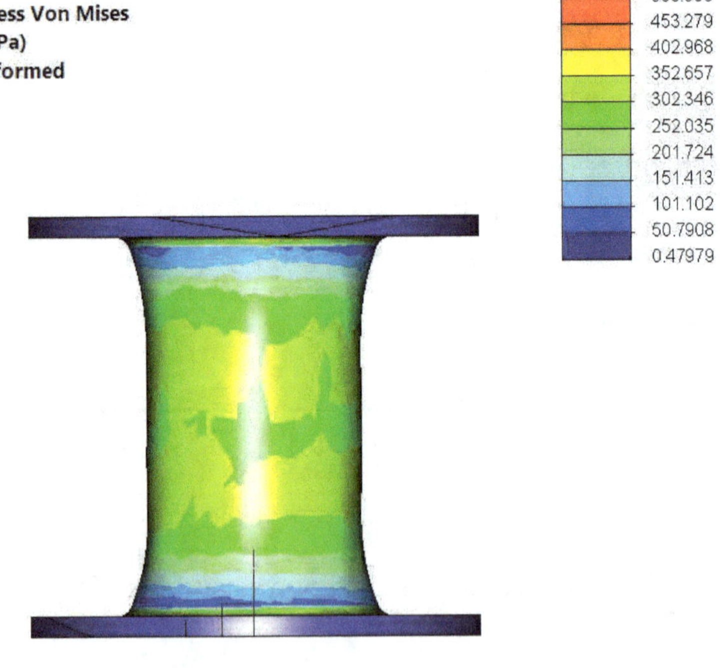

(a)

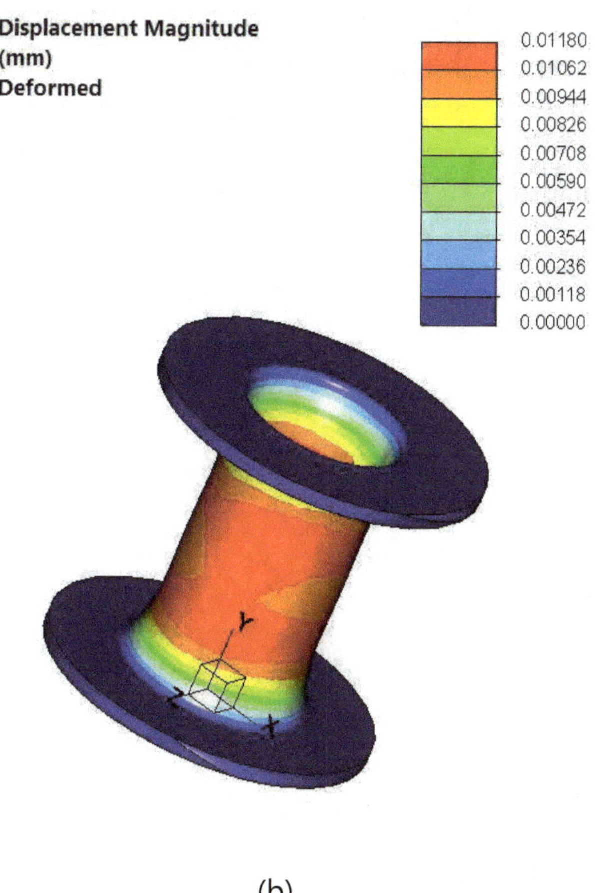

(b)

Figure 2.6-2: FEM results.

On Figure 2.6-2a, we can see the deformation of the reel, while on Figure 2.6-2b, we can see the field of displacement: it is easy to understand that reel is deformed in the same way of a cylindrical vessel under external pressure. Maximum displacement computed is of 0.013 mm. This radial displacement gives rise to incorrect solenoid operation. It is important to underline that the FEM model used herein is very simple and so results can be improved.

In order to have a comparison for FEM calculation we can apply the theory of cylindrical vessels under external pressure. Referring to [3] (chapter 3, page 84, "Recipienti cilindrici soggetti a pressione esterna in regime elastico"), the radial displacement for internal and external surfaces of a cylinder with fixed supports at its ends is:

$$Internal\ radius: u_i = -\frac{r_i}{E} p_e \frac{2\rho^2}{\rho^2 - 1}\left(1 - \frac{1}{m^2}\right) \qquad (2.6.12)$$

$$External\ radius: u_e = -\frac{r_e}{E} p_e \frac{1}{\rho^2 - 1}\left[\rho^2\left(1 - \frac{1}{m} - \frac{2}{m^2}\right) + \left(1 + \frac{1}{m}\right)\right]$$

$$(2.6.13)$$

with:

- r_i: internal radius;
- r_e: external radius;
- p_e: external pressure;
- $\rho = r_e/r_i$;
- $m = 1/v$;
- v: Poisson coefficient;
- E: Young modulus.

For the same solenoid analysed with FEM we find:

$$u_i = -0.01881mm$$
$$u_2 = -0.01776mm$$

The results have the same order of magnitude of FEM. Measurements on solenoid show that in some case the internal diameter decreases of 0.015mm showing that such a problem can exist. Therefore, it is easy to understand that a correct wire tension control is necessary to avoid malfunction of solenoid.

2.7. Observations on resistance-temperature dependency

In chapter 1, we have used the following expression:

$$R = R_0\left[1 + \alpha(T - T_0)\right]$$

Using the well-known relation between resistance and resistivity, we can write:

$$\rho = \rho_0\left[1 + \alpha(T - T_0)\right] \tag{2.7.1}$$

Expression (2.7.1) shows the linear dependency of resistivity from temperature. Now let us write (2.7.1) in a different way:

$$\rho = mT + q$$
$$m = \rho_0\alpha$$
$$q = \rho_0\left(1 - \alpha T_0\right)$$

So we can understand that, if $T_0=20°C$, we necessary have to evaluate α and ρ_0 at 20°C because m and q are constants. If $T_0=0°C$ we have to evaluate α and ρ_0 at 0°C. Therefore, reference temperature T_0 must be in accordance with wire tables we choose.

To study in deep this topic see [4].

Chapter 3

Materials and saturation

3.1. Magnetic saturation

We have model the ferromagnetic material in a linear way:

$$B = \mu H = \mu_r \mu_0 H \tag{3.1.1}$$

It is well known that this relation is not always valid but in the same time, thanks to its simplicity, it is very useful in order to size a solenoid. In effect, the relation between B and H is represented by Figure 3.1-1: this is the curve of first magnetization (Curve 1).

We can observe that, for large value of H, the magnetic induction undergoes to the so-called saturation: the curve flatten and (3.1.1) is not valid. The explanation of this behaviour stays in the nature of ferromagnetic materials. From magnetic point of view, these materials consist of an ensemble of the so-called Weiss domains. These domains, in a similar way of electrical polarization, tends to align to magnetic field direction, increasing magnetic induction value. After that, all domains are aligned, the material is saturated and behaves like air or vacuum: at that point the magnetic permeability is equal to the value in air or vacuum. The magnetic induction at saturation is called saturation induction B_S.

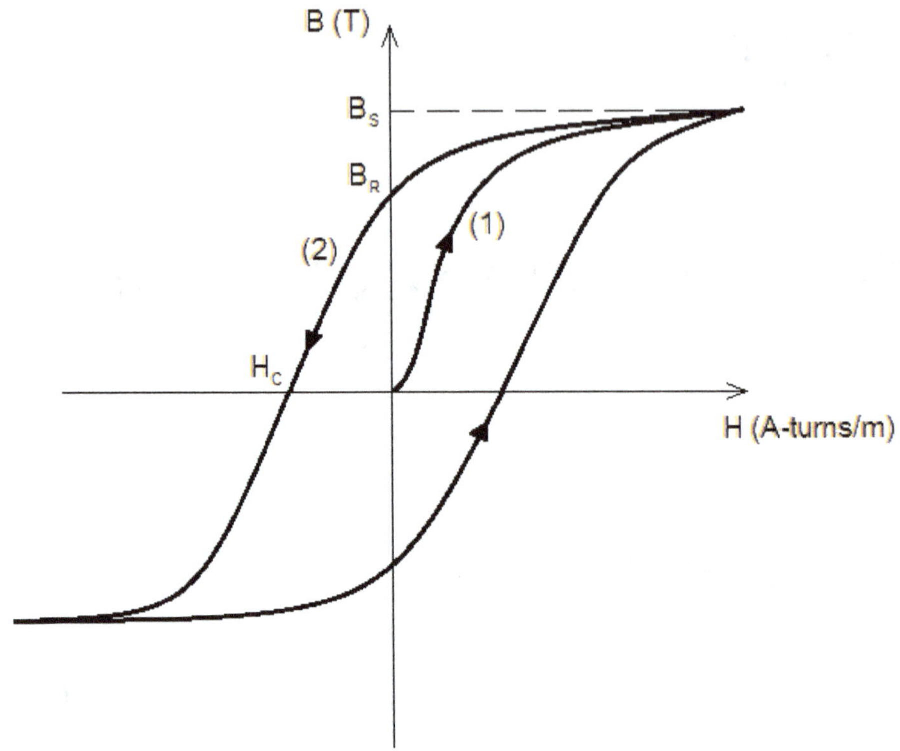

Figure 3.1-1: Magnetization curves and hysteresis.

After saturation, if we decrease the value of magnetic field H, we will see that material does not follow the curve of first magnetization but tends to keep high value of magnetic induction even for very low value of H. When $H=0$, we can see a non-zero value of B, the so-called Magnetic Remanence B_R: it means that a residual magnetization is present in the material that now is able to "produce" a magnetic action on other ferromagnetic materials. To eliminate B_R, we have to "produce" a negative value of H: the magnetic field H_C that eliminate B_R is called Coercive field. From saturation to zero-value of B, the material will follow the so-called De-magnetization curve (Curve 2).

If we keep on reducing the value of H to higher negative values, the material saturates in the same way as for positive values of H. We can find a well-known hysteretic behaviour of ferromagnetic material. The characterization of hysteresis curve is necessary for AC solenoid where H oscillates between positive and negative value. For DC devices, analysing Curves 1 and 2 is enough and it provides the possibility to compute important performance parameters.

In particular, we are interested to Soft Magnetic materials. This kind of materials has the property of being easily magnetized and de-magnetized. It means that they have a very high value of the relative magnetic permeability and a low value of Coercive force.

3.2. Magnetization curves modelling

In the following, we will use magnetization curve to compute magnetic field and magnetic force even in saturated condition. This is very important because, when air gap is very little (for electro-valves has an order of magnitude of some tenth of millimeter), solenoids saturate. In general, temperature and tension ranges are very wide, that means that solenoid can work in many different conditions that can cause saturation. The best way to take in account saturation is using experimental magnetization curves. This information is usually missing because only few materials producers provide these curves. The only data available are relative magnetic permeability μ_R, magnetic induction saturation value B_S, Magnetic Remanence B_r and Coercive Force H_C.

From this data is possible to build theoretical magnetization curves. We follow the theory discussed in [5] for soft magnetic materials. The model is called "hyperbolic model" because is based on mathematical description of curves by use of hyperbolic functions. This assumption is based on physical considerations discussed in [5]. We write the function describing Curve 1 in Figure 3.1-1 as:

$$B(H) = A_0 H + B_0 \tanh(C_0 H) \tag{3.2.1}$$

Let us write the derivative of (3.2.1):

$$\frac{dB}{dH}(H) = A_0 + B_0 C_0 \left[\operatorname{sech}(C_0 H)\right]^2 \tag{3.2.2}$$

We have to find the constants A_0, B_0 and C_0. Using the properties of hyperbolic functions, we write the following conditions:

$$\left.\frac{dB}{dH}\right|_{H=0} = A_0 + B_0 C_0 = \mu_r \mu_0$$

$$\left.\frac{dB}{dH}\right|_{H\to\infty} = A_0 = \mu_0 \tag{3.2.3}$$

$$B_0 \tanh(C_0 H)\big|_{H\to\infty} = B_0 = B_S$$

From the second and the third condition of (3.2.3), we can write:

$$C_0 = \frac{\mu_0(\mu_r - 1)}{B_S} \tag{3.2.4}$$

We can write now:

$$B(H) = \mu_0 H + B_S \tanh\left(\frac{\mu_0(\mu_r - 1)}{B_S} H\right) \tag{3.2.5}$$

Now we consider the following property of hyperbolic tangent:

$$\lim_{H\to\infty} \tanh\left(\frac{\mu_0(\mu_r - 1)}{B_S} H\right) = 1 \tag{3.2.6}$$

We can write the expression of B for values of H above saturation:

$$B(H) = \mu_0 H + B_S \tag{3.2.7}$$

This equation is very important because it shows how after saturation, the ferromagnetic materials behaves like an amagnetic material, such as air or vacuum.

The function describing the curve of de-magnetization (Curve 2) is:

$$B^-(H) = B_0 \tanh\left[c_1(H + a_1)\right] + A_0 H \tag{3.2.8}$$

We write the following conditions:

$$B^-(-H_C) = B_0 \tanh\left[c_1(-H_c + a_1)\right] - A_0 H_C = 0$$
$$B^-(0) = B_0 \tanh(c_1 a_1) = B_R \tag{3.2.9}$$

Substituting the expression for A_0 and B_0 we can find:

$$c_1 = \frac{\left|\operatorname{arctan} h\left(\dfrac{B_R}{B_S}\right) - \operatorname{arctan} h\left(\dfrac{\mu_0 H_C}{B_S}\right)\right|}{H_C} \tag{3.2.10}$$

$$a_1 = \frac{1}{c_1}\operatorname{arctan} h\left(\frac{B_R}{B_S}\right)$$

3.3. Magnetic materials

We now mention some common materials used for solenoid applications. These materials are called "Soft Magnetic materials". This kind of materials has low coercive force and high relative permeability. These characteristics provides a reduced hysteresis to the material and the possibility to be magnetized and de-magnetized very easily.

<u>Pure Iron</u>

This is one of the best soft magnetic material and it is easy to find in commerce. Its high magnetic performances are due to high iron purity: it is composed by 99.85 % of iron. This characteristic provides the material very low mechanical properties and very low corrosion resistance (no percentage of chromium). So a designer that wants to use this material should be aware of the fact that no structural function can be achieved by parts made in pure iron. Moreover, the material must be protected from corrosion and from wear, because of its very low value of hardness. Usually nickel-plating (electroless) is the solution adopted to provide resistance to wear and corrosion.

Another typical feature of this material is that, after machining operation, magnetic properties are changed. It is necessary to heat treat the part to regenerate magnetic properties by annealing. Without this operation, magnetic properties are unacceptable. Moreover, special cutting tool as well as particular cutting speed are used to produce parts.

ARMCO® pure iron is one of the most common commercially available material. Its properties depend from the type of annealing after machining operations. There are two types of heat treatment classified as "A" and "B". Both of them must be made in inert atmosphere. In Figure 3.3-1, we can see the differences between the two types of annealing.

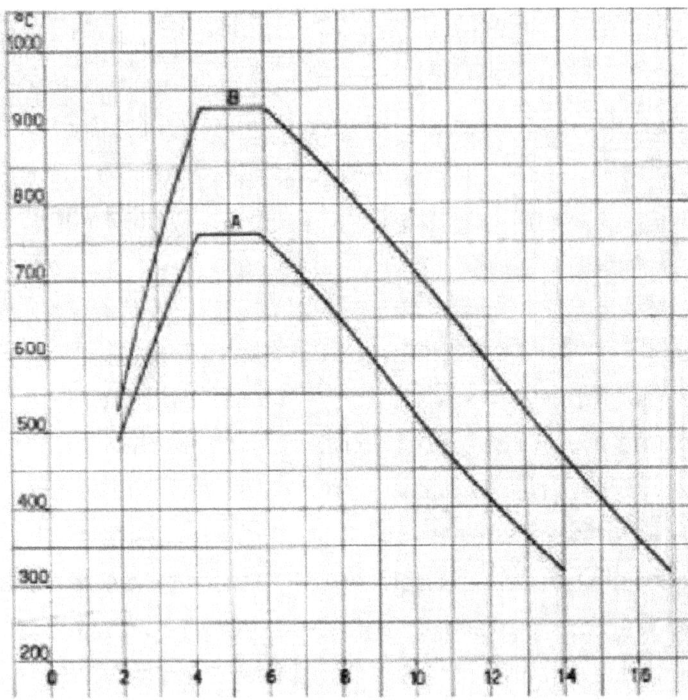

Figure 3.3-1: Annealing "A" and "B" (x axis in hours) for ARMCO®.

The physical properties for treatment "A":

Table 3.3-1: Pure Iron ARMCO® physical properties ("A" treatment)

Tensile strength, ultimate (kg/mm^2)	300
Tensile strength, yield (kg/mm^2)	190
Hardness HRB	40-50
Young modulus (kg/mm^2)	21000
Fatigue limit (kg/mm^2)	18
Density (kg/ dm^3)	7.86
Specific heat (J/(kg K))	450
Thermal conductivity (W/(m K))	73.15
Magnetic saturation (T)	2.15
Relative permeability max	3500
Relative permeability initial	200

The physical properties for treatment "B" (only differences with "A"):

Table 3.3-2: Pure Iron ARMCO® physical properties ("B" treatment)

Tensile strength, ultimate (kg/mm^2)	280
Tensile strength, yield (kg/mm^2)	140
Hardness HRB	25-35
Relative permeability max	8000
Relative permeability initial	300

The magnetic properties are shown in detail in the following figures.

The figures show that treatment "B" has higher magnetic performances than "A". In this case we can use material producer data even if is possible to apply the theory discussed above. Data herein reported are available in [6].

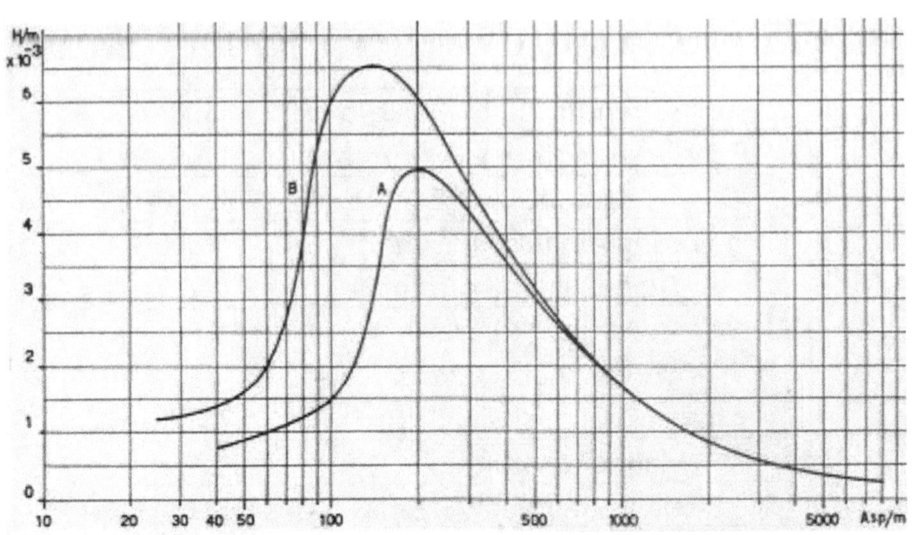

Figure 3.3-2: Magnetization curve of ARMCO®.

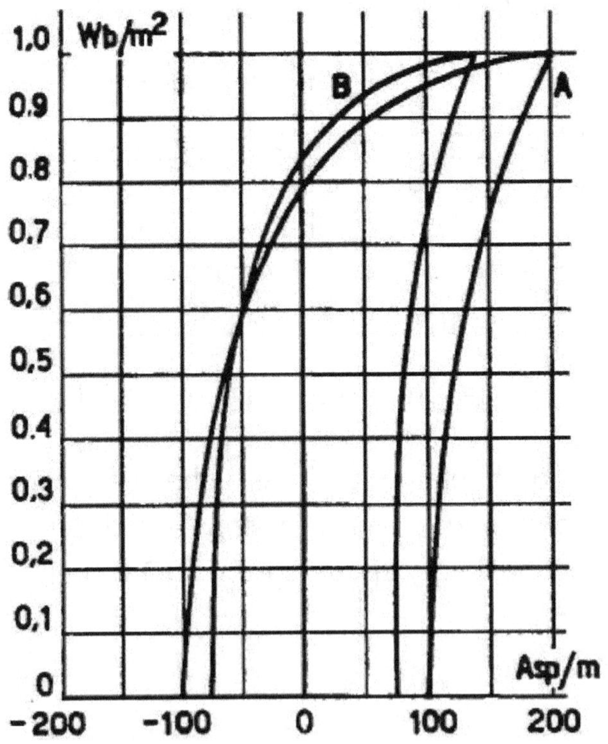

Figure 3.3-3: Hysteresis of ARMCO®.

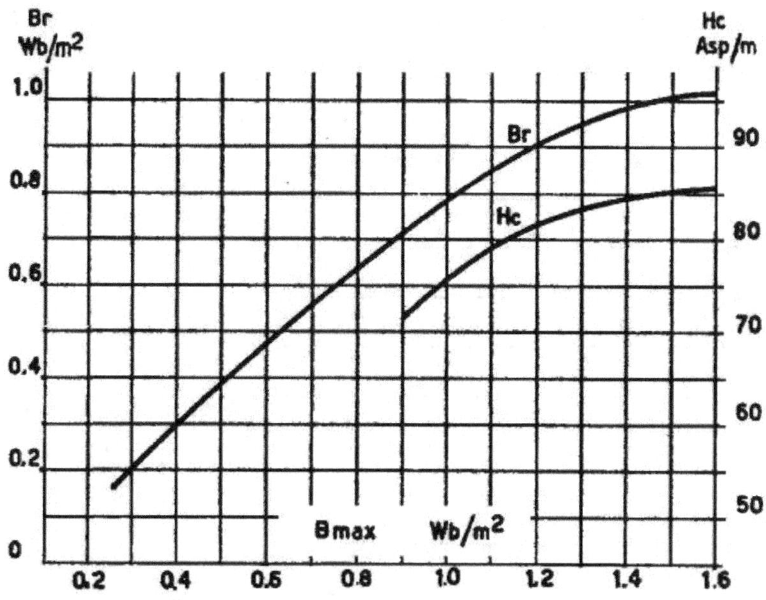

Figure 3.3-4: Magnetic Remanence and Coercive force of ARMCO®.

<u>Sandvik 1802®</u>

This material is a ferrite stainless steel. It has good mechanical properties and good corrosion resistance. It should be protected from corrosion only when required severe salt fog test.

Table 3.3-3: Sandvik 1802® magnetic properties (see [7])

Magnetic properties	As delivered	After annealing at 800°C(*)	After solution Heat treatment from 1050°C
Relative permeability	400	1000	600-1000
Saturation magnetization (T)	1.35-1.42	1.40-1.48	1.25-1.35
Remanence (T)	0.4-0.8	0.7-0.9	0.3-0.4
Coercive force (A-turns/m)	318-557	199-278.5	159-238.7

(*) Holding time 2 hours, cooling 15°C/hour to 600°C.

In this case, we have to use the above formulas to define the magnetization curves defined in Figure 3.3-5.

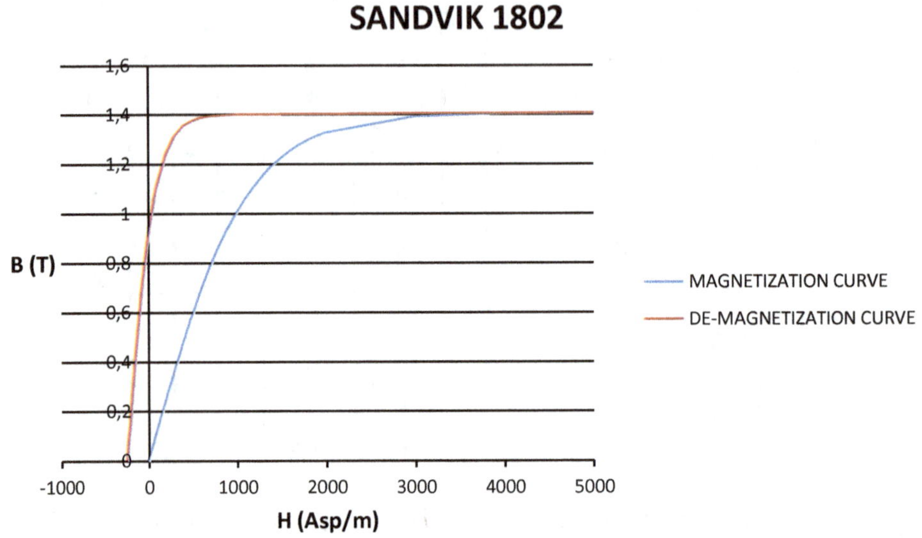

Figure 3.3-5: Magnetization curve for Sandvik 1802®.

<u>AISI 410</u>

AISI 400 series are martensitic ferromagnetic stainless steel. They does not have the performances of Sandvik 1802® but they have the great advantage to have high structural strength when heat-treated.

Table 3.3-4: AISI 410 magnetic properties (see [8])

Magnetic properties	Annealed
Relative permeability	800
Saturation magnetization (T)	1.6
Remanence (T)	0.4-0.8
Coercive force (A-turns/m)	1.2

Even in this case we have to use the above formulas to define the magnetization curves defined in Figure 3.3-6.

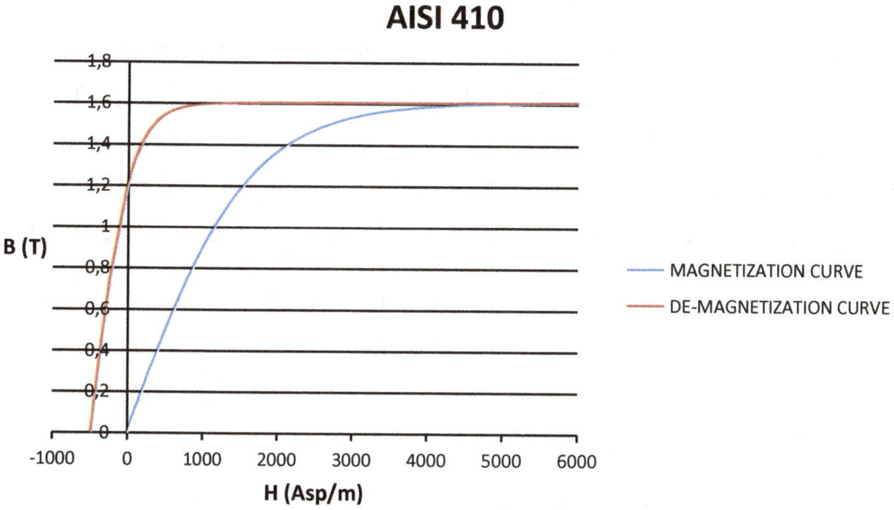

Figure 3.3-6: Magnetization curve for AISI 410.

3.4. Physical aspects of magnetic materials

Now we will focus briefly on some particular aspect to take in account in solenoid design regarding ferromagnetic materials.

<u>Annealing</u>

When materials are machined and worked, they lower their magnetic performances because of residual stresses. The process of annealing is very important because it restores magnetic properties of materials. In order to better understand this process, we analyse the Pure Iron ARMCO®. Heat treatment classified as "A" consists only in a regeneration of magnetic properties, while the heat treatment classified as "B" provides an increase of the crystal grain and so it increase magnetic properties. Therefore, it is very important to heat treat parts only when machining and/or cold working process are ended.

Corrosion resistance

It is well known that a high percentage of chromium increases corrosion resistance and that steels with a composition of 18% of chromium or above are called stainless steel. In effect, all stainless steel parts are surface finished with a process of passivation that provides a total corrosion resistance. In the case of solenoid devices, ferromagnetic parts are often nickel plated, usually with the electroless nickel-plating process, which is more uniform than electrolytic process.

The high percentage of chromium guarantees a good corrosion resistance but on the other side decreases magnetic properties: on Figure 3.4-1, it has shown the dependency of saturation magnetization from chromium percentage.

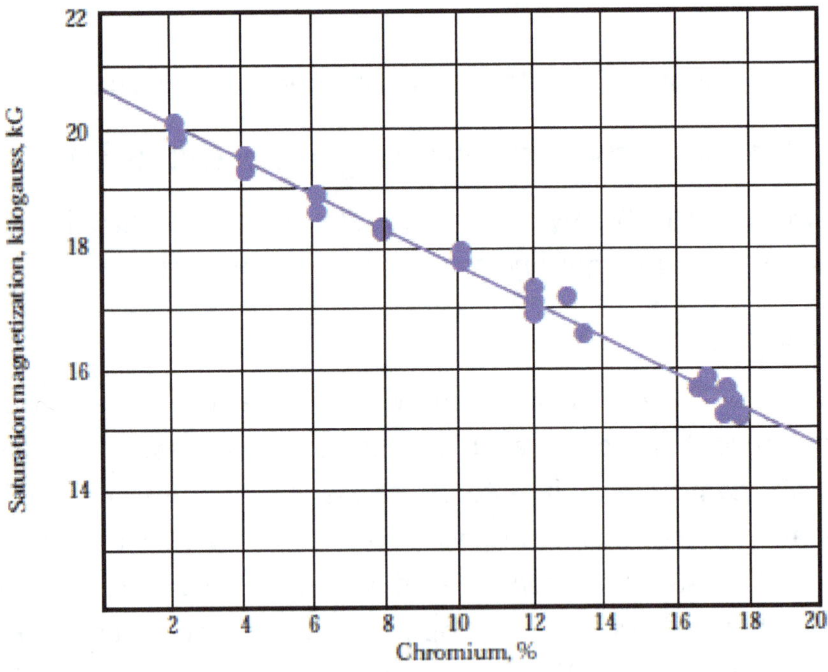

Figure 3.4-1: Saturation magnetization Vs. chromium perc. (from [9]).

3.5. Amagnetic materials

It is also important to briefly analyse amagnetic materials properties. Usually these materials are austenitic AISI 300 series. Looking at Figure 1.2-1 it is very important that some parts are made absolutely of amagnetic steel. The reason stays in the necessity of create a perfect magnetic circuit were magnetic induction can flow without been deviated or saturated. In particular, in Figure 1.2-1, we can see that the top of the solenoid is amagnetic: if it is ferromagnetic, magnetic induction will flow even in this part and keeper wold be attracted by the top of the solenoid.

These materials are easily weldable and have structural strength similar to ferromagnetic material. Their most important characteristic is to have relative magnetic permeability approximately near to the unity value. In effect, a perfect amagnetic material is characterized by this value of μ_r. This behaviour is ideal because even materials considered amagnetic can be attracted by a magnet. This fact does not mean that their not suitable in solenoid application, because in that devices they are installed near to parts with higher relative permeability of two or three order of magnitude than amagnetic materials. Therefore, magnetic flux is practically not affected by this behaviour. In [10] we can find Figure 3.5-1.

We cite [11] to comment Figure 3.5-1: "in solubilised state, austenitic steel is 'non magnetic' whilst dissolved but,..., they develop magnetic properties when they are cold formed, because this operation is capable of transforming a part of austenite into martensite". We also notice that AISI 316 is not affected by this problem and so it represents the best choice in solenoid design.

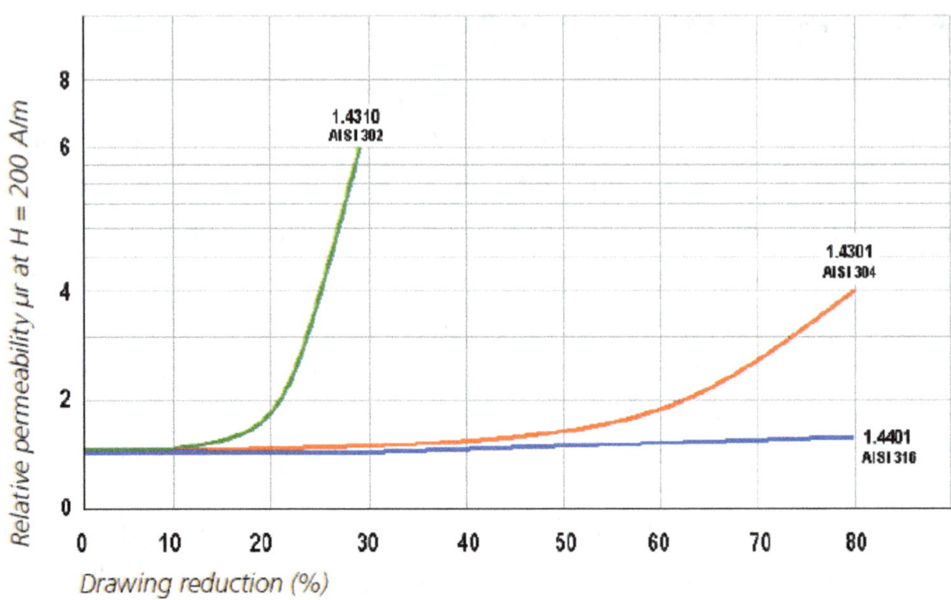

Figure 3.5-1: Relative magnetic permeability for amagnetic materials (from [10]).

Chapter 4

Complete Calculation

4.1. Saturation computation

In the previous chapters, we have written the Ampère Law assuming that the dependency of magnetic induction from magnetic field is linear. We can write this law without this assumption in the following way:

$$H_m l_m + H_0 l_0 = NI \qquad (4.1.1)$$

where H_m is the magnetic field assumed as constant in the ferromagnetic material, while H_0 is the magnetic field in air gap. Therefore, we can write:

$$H_m l_m + \frac{B_0 l_0}{\mu_0} = NI \rightarrow H_m\left(\frac{\phi}{S}\right)l_m + \frac{l_0}{S\mu_0}\phi = NI \qquad (4.1.2)$$

In this equation we can see the function H_m describing the inverse function of $B=B(H)$. This function cannot be expressed in an explicit form, even in the case the magnetic material is expressed with the formulas described in chapter 3. So, if the input is the current I, the problem can be solved by numerical methods.

We can easily solve (4.1.2) if we use a different approach:

- The input variable is H and from magnetization curve we know the corresponding value of B;
- Using the value of the crossing area S of magnetic circuit we can compute the magnetic flux Φ;
- From (4.1.2), we compute the corresponding value of I.

In this way, we can know easily the function $B=B(I)$. Then, we use the already explained formula (1.2.13) we compute the force:

$$F = \frac{B^2 S}{2\mu_0} \qquad (1.2.13)$$

We obtain a graph as in Figure 4.1-1.

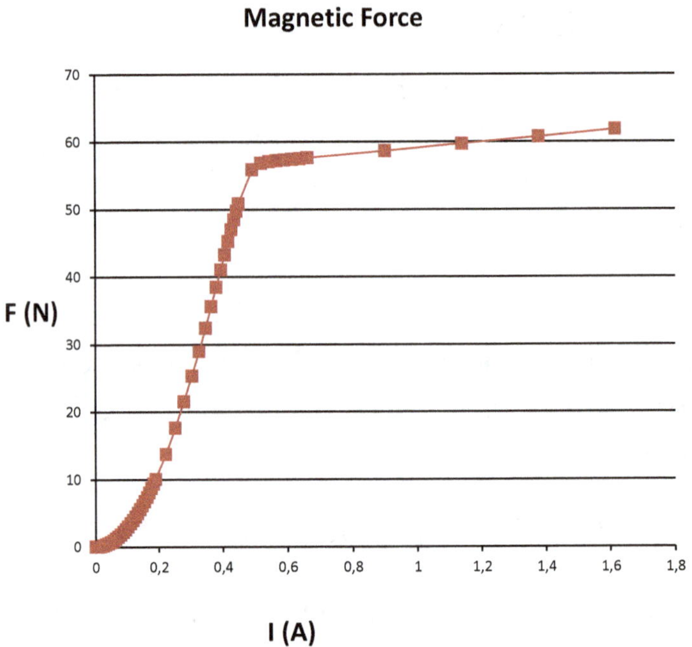

Figure 4.1-1: Magnetic force Vs. Current.

In Figure 4.1-1, we can notice how magnetic force initially depends from current in a parabolic way (see (1.2.17)). When the magnetic material saturates, even force saturates. In this condition, we can try to understand the behaviour of force.

At first, we assume that all flux lines are contained in the ferromagnetic material. This a strong assumption because, after saturation, we have understood from §3.1 that the ferromagnetic material behaves as air. Therefore, after saturation, flux lines will start to get out from magnetic circuit. However, this assumption is valid at the beginning of saturation process.

We write the magnetic induction in ferromagnetic material using (3.2.7) as:

$$B = \mu_0 H_m + B_S \tag{4.1.3}$$

The magnetic field:

$$H_m = \frac{B - B_S}{\mu_0} \tag{4.1.4}$$

We write the Ampère Law as:

$$H_m l_m + \frac{B l_0}{\mu_0} = NI \tag{4.1.5}$$

We substitute (4.1.4) in (4.1.5) and we express the magnetic induction:

$$B = \frac{N\mu_0}{l} I + \frac{B_S l_m}{l} \tag{4.1.6}$$

where $l = l_m + l_0$. Now we can write the derivative of B respect to I in the saturation zone:

$$\frac{dB}{dI} = \frac{N\mu_0}{l} \tag{4.1.7}$$

Now we write the force as (1.2.13) and we compute the derivative of force respect to current in the saturation zone using (4.1.7):

$$\frac{dF}{dI} = \frac{S}{2\mu_0} 2B \frac{dB}{dI} = \frac{S}{\mu_0} B \frac{dB}{dI} = \frac{BSN}{l} \tag{4.1.8}$$

Substituting (4.1.6) in (4.1.8) we find:

$$\frac{dF}{dI} = A + BI$$

$$A = \frac{B_S NSl_m}{l^2} \qquad (4.1.9)$$

$$B = \frac{SN^2 \mu_0}{l^2}$$

From (4.1.9), we see that magnetic force has a parabolic dependency from current. However if we substitute typical values for a solenoid we find that:

$$A = 4.246 \frac{N}{A}$$

$$B = 0.1603 \frac{N}{A^2}$$

We can understand from this result that in practice the dependency is almost linear.

The computation of force in saturation can be made with another method called Coenergy Method. First of all, we define the Linkage flux as:

$$\lambda = N\varphi \qquad (4.1.10)$$

Coenergy is defined as:

$$W_{CO} = \int_0^{I_A} \lambda dI \qquad (4.1.11)$$

The physical meaning is shown in Figure 4.1-2 (refer to [12] for further discussions).

It is possible to demonstrate that magnetic force can be expressed as (magnetic field is conservative):

$$F = \frac{\partial W_{CO}}{\partial x}\bigg|_{I_A} \qquad (4.1.12)$$

where ∂x represents the infinitesimal variation of air gap. This method is more difficult to apply, but it is equivalent to the one applied here using Maxwell Stress Tensor (see Figure 4.1-3).

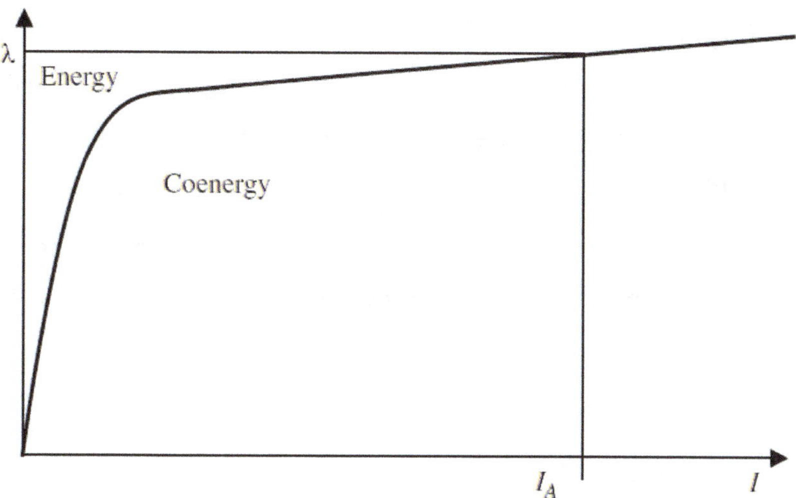

Figure 4.1-2: Flux linkage and Coenergy (from [12]).

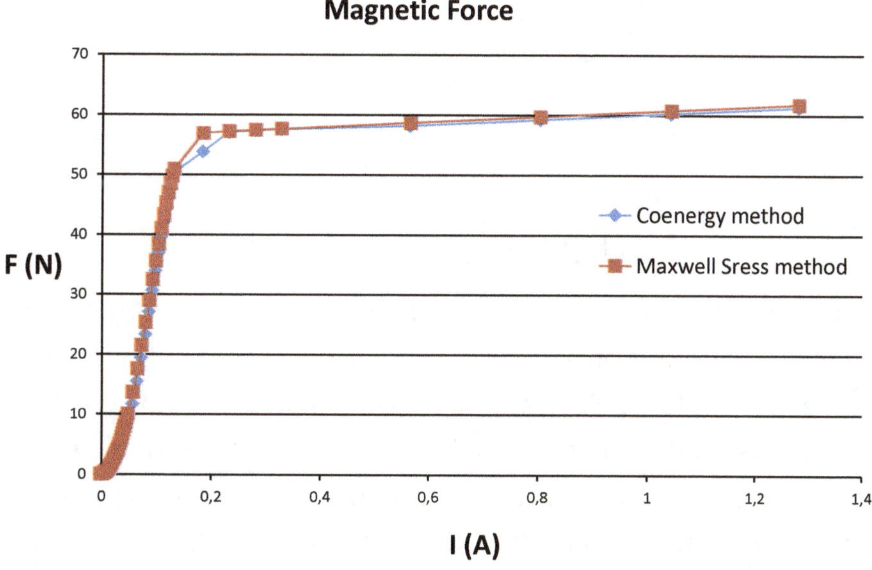

Figure 4.1-3: Coenergy method Vs. Maxwell Stress method.

As we can see from Figure 4.1-3, the method are equivalent. The only difference we can notice is in the transition zone where the integration rule used for Coenergy method is less accurate.

4.2. Fringing effect

In order to compute accurately solenoid performances, it is important to take into account fringing effect. First of all, we have to understand what is fringing.

In the proximity of the air gap magnetic flux tends to leave the keeper (or the core it depends from the direction of flux) as shown in Figure 4.2-1.

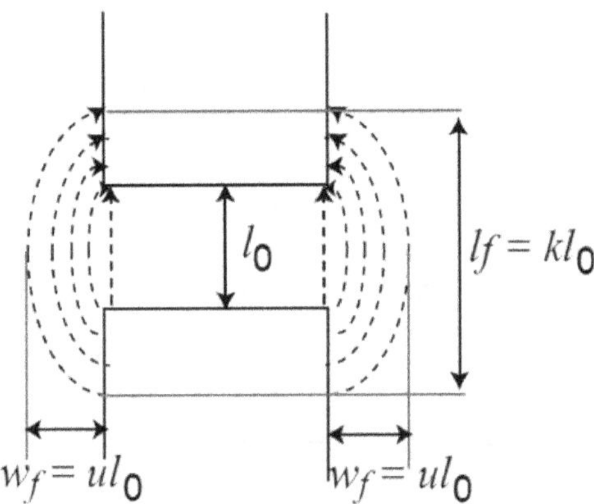

Figure 4.2-1: Fringing effect.

As we can see from Figure 4.2-1, not all flux lines are normal to keeper surface. This condition influences the value of magnetic induction on the keeper surface. Basically, the total magnetic flux produced by windings is split in two near the air gap. A fraction of flux passes through keeper surface and to air gap; the other fraction passes on the sides of the theoretical air gap. It means that only the fraction passing through the air gap gives his contribution to magnetic force. It is easy to understand that this effect is much more important when air gap increases.

To take in account fringing effect, following the analysis discussed in chapter 1, we define another magnetic circuit model shown in Figure 4.2-2.

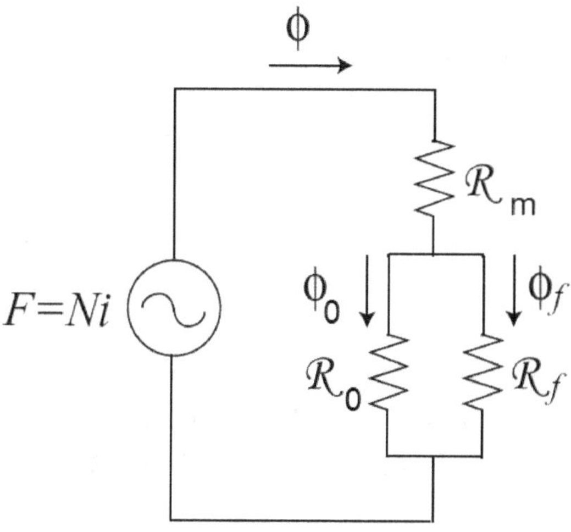

Figure 4.2-2: Magnetic circuit definition for fringing effect.

We define the Permeance as:

$$\wp = \frac{1}{\mathcal{R}} \qquad (4.2.1)$$

The Permeance of the fringing zone is:

$$\wp_{0f} = \wp_0 + \wp_f = \wp_0\left(1 + \frac{\wp_f}{\wp_0}\right) = \frac{\mu_0 S}{l_0} + \frac{\mu_0 A_f}{l_f} = \frac{\mu_0 S F_f}{l_0} \qquad (4.2.2)$$

where we have defined the Fringing Factor:

$$F_f = \frac{\wp_{0f}}{\wp_0} = 1 + \frac{\wp_f}{\wp_0} = 1 + \frac{A_f l_0}{S l_f} = 1 + \frac{A_f}{S k} \qquad (4.2.3)$$

where l_f is the length of the fringing zone as shown in Figure 4.2-1 and k is:

$$k = \frac{l_f}{l_0}$$

(4.2.4)

Considering Figure 4.2-2, the total reluctance of the solenoid before saturation is:

$$\Re = \Re_m + \frac{1}{\wp_{0f}} = \frac{l_0}{\mu_0 S}\left(\frac{l_m}{\mu_r l_0} + \frac{1}{F_f}\right)$$

(4.2.5)

In general, we can write (because $l_m/\mu_r l_0 << 1$)

$$H_m\left(\frac{\varphi}{S}\right)l_m + \frac{l_0}{\mu_0 S F_f}\varphi = NI$$

(4.2.6)

With reference to Figure 4.2-1, we define the following parameter:

$$u = \frac{w_f}{l_0}$$

(4.2.7)

with w_f the radial amplitude of the fringing zone. Now we write for tubular solenoid S as (1.2.14) and A_f (see [13]) as:

$$A_f = \frac{\pi(D_1 + 2ul_0)^2}{4} - \frac{\pi D_1^2}{4} = \pi u l_0(D_1 + ul_0)$$

(4.2.8)

Substituting (1.2.14) and (4.2.8) in (4.2.3) we can write the Fringing Factor:

$$F_f = 1 + \frac{4ul_0(D_1 + ul_0)}{(D_1^2 - D_{STEL}^2)k}$$

(4.2.9)

To calculate magnetic force we first need to compute the flux φ_0. We write the following system of equations:

$$\begin{cases} \varphi = \varphi_0 + \varphi_f \\ \Re_0\varphi_0 = \Re_f\varphi_f \end{cases} \rightarrow \varphi_0 = \frac{\Re_f}{\Re_0 + \Re_f}\varphi$$

(4.2.10)

The magnetic induction in the air gap is:

$$B_0 = \frac{\Re_f}{\Re_0 + \Re_f}\frac{\varphi}{S} = \frac{1}{F_f}\frac{\varphi}{S}$$

(4.2.11)

where φ is computed solving (4.2.6). Then we can compute the force:

$$F = \frac{B_0^2 S}{2\mu_0} \tag{4.2.12}$$

As a general rule, we will use the model with the assumption that u=1, k=1. To verify the accuracy of the model see chapter 5, regarding Finite Element Method for magnetic problems.

4.3. Inductance Calculation

Computation of inductance is very important because it is a parameter that influence transitory response of the solenoid. First of all, we can compute the inductance for a solenoid with magnetic material non saturated. This computation shows the parameters who affects inductance.

From the definition of inductance:

$$V_i = L\frac{dI}{dt} \tag{4.3.1}$$

where V_i is the electro-motive force induced by inductance due to current variation. The law of Faraday-Neumann-Lenz for a coil composed by N windings is:

$$V_i = \frac{d(N\varphi)}{dt} \tag{4.3.2}$$

Equating (4.3.1) and (4.3.2) and integrating we find:

$$L = \frac{N\varphi}{I} \tag{4.3.3}$$

If we assume a linear dependency of B from H, we can write the Hopkinson Law and solve for φ:

$$\varphi = \frac{NI}{\mathfrak{R}} \tag{4.3.4}$$

Substituting (4.3.4) in (4.3.3) we find:

$$L = \frac{N^2}{\Re}$$

(4.3.5)

If we approximate the reluctance with the reluctance of the air gap, we can write:

$$L \approx \frac{N^2}{\Re_{GAP}} = \frac{N^2 S \mu_0}{l_0}$$

(4.3.6)

For a typical electro-valve solenoid, we can find a typical value of 0.5 H (Henries).

In the discussion above, we have assumed that the dependency of magnetic induction B from magnetic field H is linear. If we consider the real dependency of B from H, we shall use the formula (4.2.6) to compute the magnetic flux φ. Then we can compute the linkage defined as $\lambda = N\varphi$ and plot the graph shown in Figure 4.3-1.

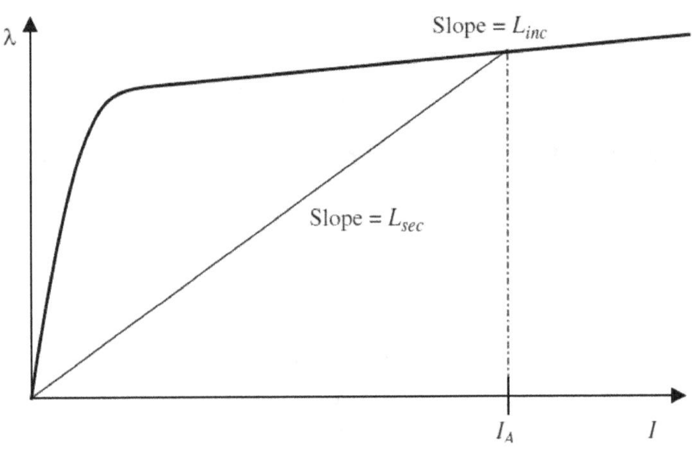

Figure 4.3-1: Linkage and inductance (from [12]).

Another equivalent definition for inductance is the following:

$$L = \frac{\lambda}{I}$$

(4.3.7)

This definition is unique when the dependency is linear. In saturation, we can define two kind of inductance: Secant Inductance and Incremental Inductance, both shown in Figure 4.3-1. The first one is defined as (4.3.7). The second is defined as:

$$L_{inc} = \frac{\partial \lambda}{\partial I} \qquad (4.3.8)$$

If we apply the (4.3.7), we can find the graph shown in Figure 4.3-2.

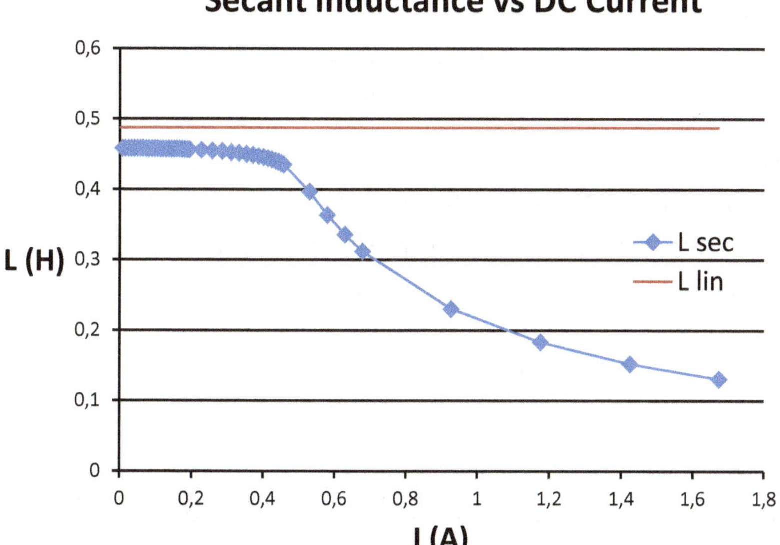

Figure 4.3-2: Secant Inductance Vs. Current.

4.4. Leakage flux computation

In the previous paragraph, we have assumed that all flux lines are contained in the ferromagnetic material. This is untrue and leads to an under estimation of magnetic force produced by the solenoid. This consideration may seem to be not intuitive because leakage flux is not contained in the ferromagnetic material and so it is not subjected to the advantage of high permeability of ferromagnetic material. However, leakage flux, in the case of a good solenoid design, exists only when ferromagnetic material saturation is occurred. In this condition, the relative permeability of ferromagnetic material is equal to the unity and the material has an amagnetic behaviour.

For understanding better these concepts, we try to explain the physical phenomena of saturation. The constitutive equation $B=B(H)$ has been written in the form of (3.2.1) or from experimental data and used in (4.1.2) and (4.2.6) to solve the magnetic circuit. Now we write this relation in a general form:

$$B_m = \mu_m\left(H_m\right)H_m \rightarrow \frac{B_m}{\mu_m\left(B_m\right)} = H_m \qquad (4.4.1)$$

Using the equation (4.4.1) in the Ampère Law, we obtain a generalized Hopkinson Law and a magnetic circuit that is composed by ordinary constant reluctances and a non-linear reluctance of the ferromagnetic material, which is:

$$\Re_m = \frac{l_m}{\mu_m\left(B_m\right)S} = \frac{l_m}{\mu_m\left(\phi_m\right)S} \qquad (4.4.2)$$

Now let us notice that after saturation some magnetic flux lines get out from ferromagnetic material and lie on air. This flux lines follow a path on air that depends from current intensity and solenoid geometry. However, if the installation of the solenoid is correct, all these flux lines enter again in the ferromagnetic material. So they can be seen in the same way of fringing flux lines. Therefore, it is clear now that the general form of the solenoid magnetic circuit is represented by Figure 4.4-1.

In Figure 4.4-1, we can see the leakage flux, called φ_a ("air") and the corresponding reluctance $\Re_a$. This parameter can be considered constant with current and is defined as:

$$\Re_a = \frac{k_a}{\mu_0} \qquad (4.4.3)$$

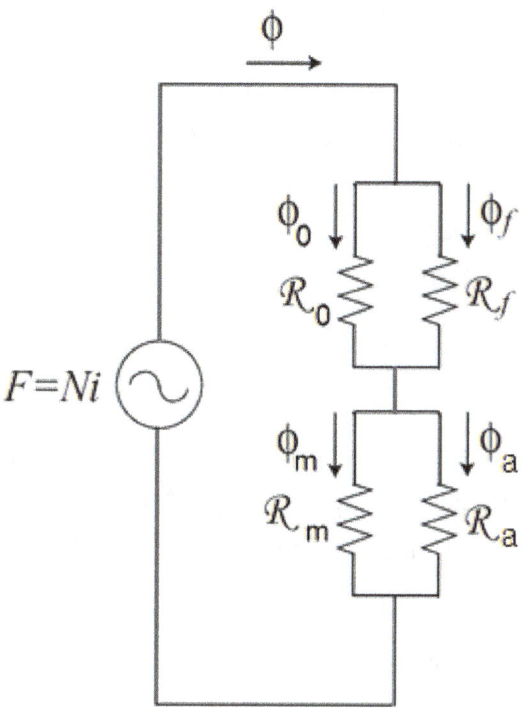

Figure 4.4-1: Solenoid general magnetic circuit representation.

where k_a is a constant that, from FEM results comparison, is assumed to be:

$$k_a = 0.1mm^{-1} \tag{4.4.4}$$

Now let us focus the attention on the magnetic circuit in Figure 4.4-1. We can define the equivalent reluctance of the parallel between magnetic material and air path of the leakage flux:

$$\mathcal{R}_{am} = \frac{\mathcal{R}_m(B_m)\mathcal{R}_a}{\mathcal{R}_m(B_m)+\mathcal{R}_a} \tag{4.4.5}$$

When magnetic induction is lower than saturation value, we have that $R_m \ll R_a$ and so:

$$\mathfrak{R}_{am} = \frac{\mathfrak{R}_m \mathfrak{R}_a}{\mathfrak{R}_m + \mathfrak{R}_a} \approx \frac{\mathfrak{R}_m \mathfrak{R}_a}{\mathfrak{R}_a} = \mathfrak{R}_m \tag{4.4.6}$$

In this case, magnetic flux is fully contained in the ferromagnetic material. When the ferromagnetic material reaches saturation the magnetic permeability $\mu_m(B_m)$ starts to lower until $\mathfrak{R}_m$ assumes a value of the same order of magnitude of $\mathfrak{R}_a$. In this case, we find that $\mathfrak{R}_m$ is not negligible in (4.4.5) and so, from the definition of $\mathfrak{R}_{am}$ we find that:

$$\mathfrak{R}_{am} = \frac{\mathfrak{R}_m \mathfrak{R}_a}{\mathfrak{R}_m + \mathfrak{R}_a} < \mathfrak{R}_m \tag{4.4.7}$$

So it is easy to understand that with respect of the computation in §4.1 and §4.2, where we consider only $\mathfrak{R}_m$ (in a different but equivalent way), we obtain a higher total magnetic flux φ, and so a higher force.

From the computation point of view we have the same difficulties of §4.1 and §4.2 because of saturation. We have to solve the following equation with respect to φ:

$$\left(\mathfrak{R}_{am}(\phi_m) + \mathfrak{R}_{0f}\right)\phi = NI \tag{4.4.8}$$

We can solve (4.4.8) with a simple approach:

1. We assume B_m and so $\varphi_m = B_m S$;
2. From the relation $\mu_m = \mu_m(B_m)$ we compute $\mathfrak{R}_m$;
3. From the following system (parallel between $\mathfrak{R}_a$ and $\mathfrak{R}_m$) we calculate φ:

$$\begin{cases} \phi_m + \phi_a = \phi \\ \mathfrak{R}_m \phi_m = \mathfrak{R}_a \phi_a \end{cases} \rightarrow \phi = \phi_m \left(1 + \frac{\mathfrak{R}_m}{\mathfrak{R}_a}\right) \tag{4.4.9}$$

4. Than we compute the current I.

So we obtain the relation $\varphi = \varphi(I)$. Using the formulas (4.2.10), (4.2.11) and (4.2.12), we than obtain the magnetic force. The comparison between FEM and the theoretical computation explained in this paragraph and in §4.2 is shown in Figure 4.4-2.

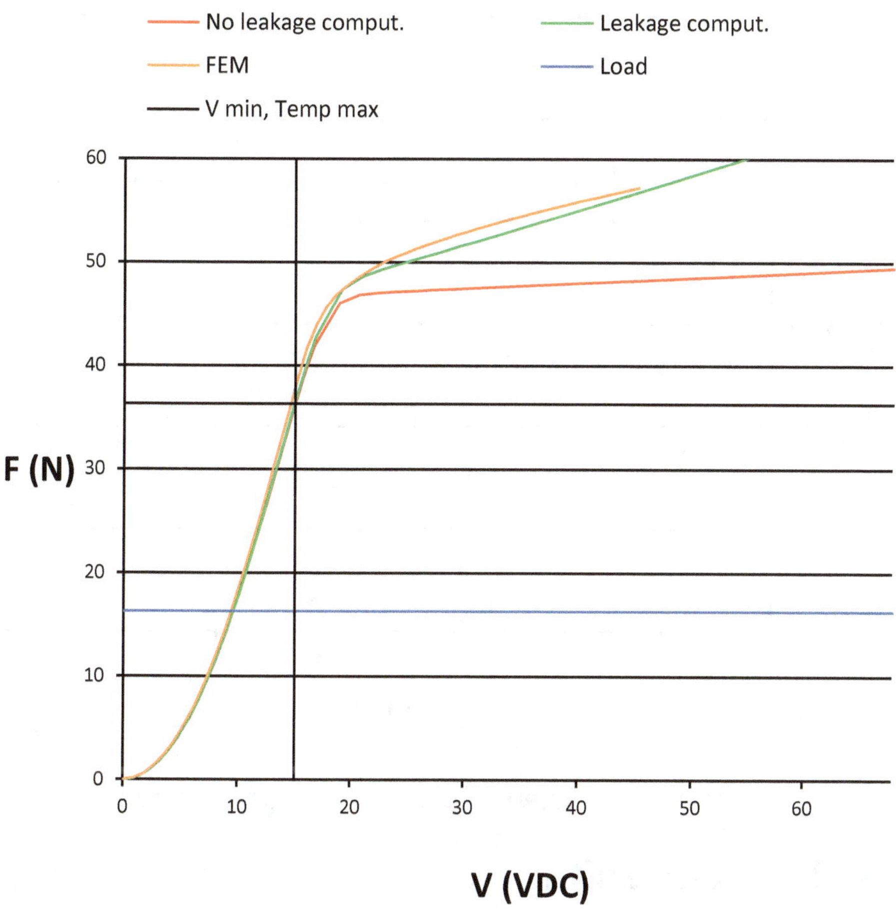

Figure 4.4-2: Solenoid force computation.

If we consider the leakage flux phenomena, we then can follow the same procedure already made to compute Secant Inductance. We obtain the results shown in Figure 4.4-3.

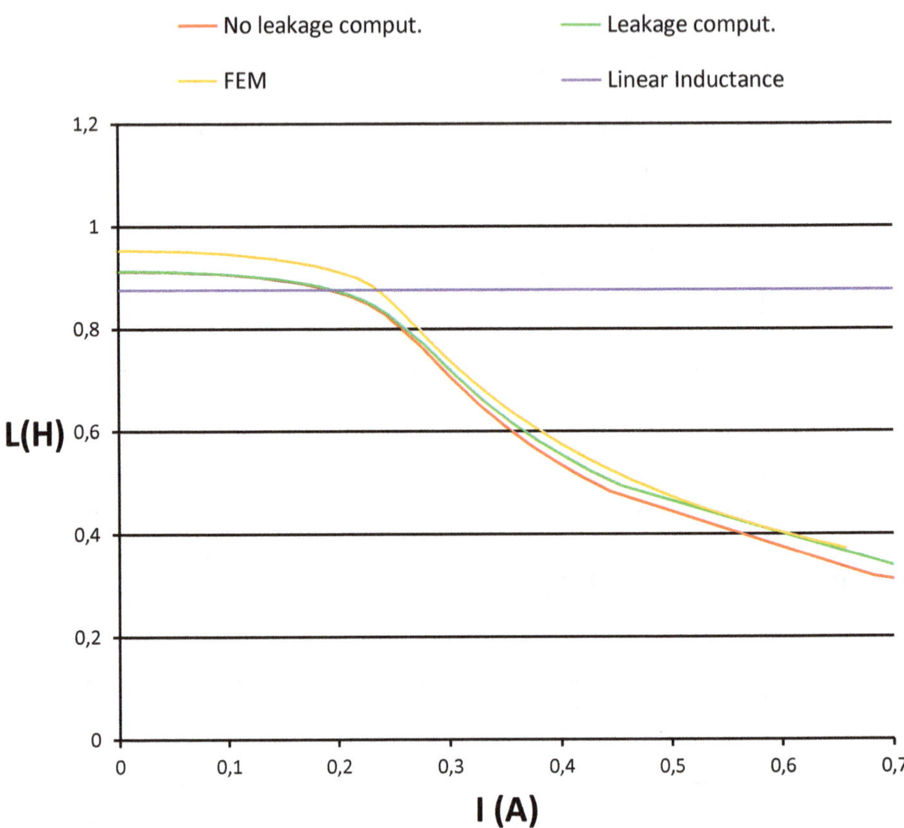

Figure 4.4-3: Solenoid Inductance computation.

4.5. Dropout Voltage

After solenoid excitation, if we slowly reduce tension or current, we notice a different behaviour of solenoid. If we recall chapter 3, we can easily understand that the reason stays in magnetic material properties: magnetic material follows the de-magnetization curve (curve 2 of Figure 3.1-1). Therefore, if we compare solenoid status for the two curves of Figure 3.1-1, solenoid will produce different forces at the same current because of residual magnetization.

For understanding better this concept, let us take a look at Figure 4.5-1. In Figure 4.5-1 some curves are represented: magnetization curve, de-magnetization curve and some dotted straight lines.

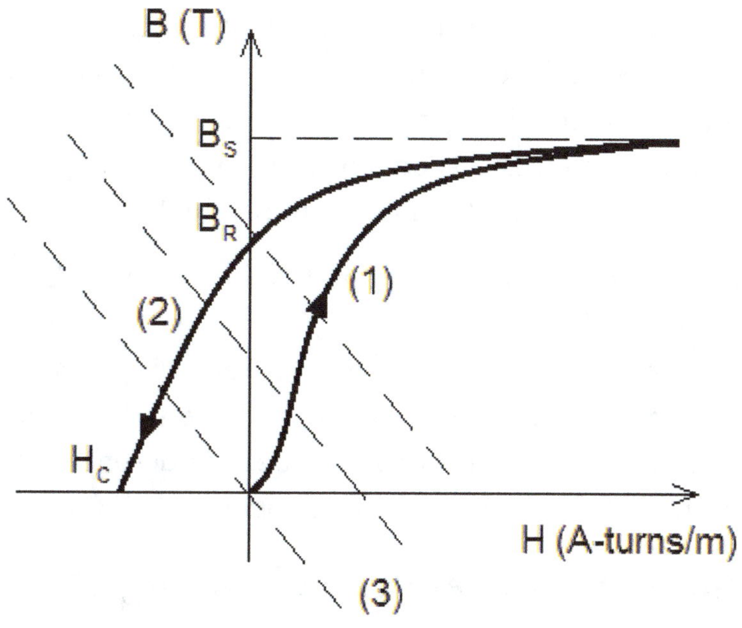

Figure 4.5-1: Dropout voltage.

These straight lines (curves 3) are obtained in the following way. We apply the Ampère Law and we explicit the magnetic induction:

$$H_m l_m + \frac{B}{\mu_0} l_0 = NI \tag{4.5.1}$$

$$B = \frac{NI\mu_0}{l_0} - \left(\frac{l_m \mu_0}{l_0}\right) H_m \tag{4.5.2}$$

Formula (4.5.2) is the equation of curves (3) and it gives the point of intersection with de-magnetization curve. From (4.5.2), we can easily understand that for positive values of current we can find a negative magnetic field H and a positive magnetic induction B.

The dropout voltage is the tension for which the magnetic force is equal to external load. To find this value, we have to solve (4.5.1) respect to B and then compute the force with (1.2.13). In (4.5.1) we have to use for H_m the de-magnetization curve. This procedure does not give an accurate result but an order of magnitude for dropout voltage, because we have assumed that all points of magnetic material are equally saturated and so that they de-magnetize in the same way. This is untrue but gives the possibility to obtain easily an order of magnitude for computing this important parameter.

Dropout voltage is very important because it describes the way solenoid closes. If, in solenoid design, we find a too low dropout voltage it is possible that it is not able to de-energize. This event is very dangerous because there can always be voltage spike in the electrical system and, if the solenoid pilot valve does not de-energize, even the valve on which is installed is not able to operate.

4.6. **Notes on tolerances and plating effects**

We want to finish this chapter considering that real parts that compose the magnetic circuit are installed with plays due to tolerances. Moreover, ferromagnetic materials are often nickel-plated in order to protect parts from wear and corrosion. Nickel-plated deposition is amagnetic. Therefore, nickel-plated layers and plays between parts form gaps of amagnetic materials that can influence total magnetic flux. Therefore, it is necessary to take into account the corresponding reluctances.

In order to do that, we can evaluate the entity of gaps length l_i and gaps flux crossing surfaces S_i. Therefore, we can write the total parasite reluctance as:

$$\Re_p = \sum_i \Re_i = \sum_i \frac{l_i}{\mu_0 S_i} \qquad (4.6.1)$$

The simplest way to take in account this parasite reluctance is to evaluate an air gap correction:

$$\mathfrak{R}_p = \frac{l_{eq}}{\mu_0 S} \rightarrow l_{eq} = \mathfrak{R}_p \mu_0 S \tag{4.6.2}$$

This equivalent air gap shall be added to the air gap we have considered in the previous paragraphs.

Chapter 5

Finite Element Method

5.1. FEM introduction

We have shown how it is possible to size a solenoid and to compute its performances with methods based on magnetic circuit theory. These methods are called "Reluctance Methods". However, there is another powerful method to compute these performances, which is the Finite Element Method. It consists in solving the Maxwell equations in all the points of the solenoid ad its surroundings. Therefore, it is necessary to know and represent the geometry of the solenoid and the relevant magnetic properties of the materials of the solution field. Then, it is necessary to numerical solve the Maxwell equations. This operation is typically done with FEM.

5.2. Physical background

Now we discuss the equation to be solved with FEM method. First of all, we write some important vectorial identities:

$$\nabla \cdot (\nabla \times V) = 0 \qquad (5.2.1)$$

$$\nabla \times (\nabla V) = 0 \qquad (5.2.2)$$

$$\nabla \cdot (\nabla V) = \nabla^2 V \qquad (5.2.3)$$

$$\nabla \times (\nabla \times V) = -\nabla^2 V + \nabla(\nabla \cdot V) \tag{5.2.4}$$

In magnetostatics, we have to solve in every point of the field the following Maxwell equations:

$$\nabla \cdot B = 0 \tag{5.2.5}$$

$$\nabla \times H = j \tag{5.2.6}$$

Formula (5.2.5) is the Magnetic Gauss Law while (5.2.6) is the Ampère Law. Now it is convenient to reduce (5.2.5) and (5.2.6) in a unique equation. We use the vectorial identity (5.2.1) and we write:

$$B = \nabla \times A \tag{5.2.7}$$

Thanks to (5.2.1), the assumption (5.2.7) solves equation (5.2.5). Now we have to notice that vector field A, called magnetic vector potential, is not uniquely defined by (5.2.7). In effect, let us define another field A_1 as:

$$A_1 = A + \nabla \phi \tag{5.2.8}$$

where ϕ is a scalar function. We substitute A_1 in (5.2.7) using the vectorial identity (5.2.2):

$$\nabla \times A_1 = \nabla \times A + \nabla \times (\nabla \phi) = \nabla \times A \tag{5.2.9}$$

Equation (5.2.9) shows how also the vectorial field A_1 solves the equation (5.2.7): it means that the vectorial field A is not uniquely defined and it is necessary another condition. This fact is confirmed by the Helmholtz theorem, also known as the fundamental theorem of vector calculus, which states that any sufficiently smooth, rapidly decaying vector field in three dimensions is completely defined by the knowledge of its curl and its divergence. In effect, if we apply the operator of divergence in (5.2.8) and we use the vectorial identity (5.2.3), we find the following equation:

$$\nabla \cdot A_1 = \nabla \cdot A + \nabla^2 \phi \tag{5.2.10}$$

It means that the field A_1 and A have different divergences but the curl of both of them gives the same magnetic induction field B and so the same magnetic field H thanks to the constitutive equation:

$$B = \mu H \tag{5.2.11}$$

Therefore, we can choose the divergence condition in the most convenient way. This choice comes from equation (5.2.6). We substitute (5.2.7) and (5.2.11) in (5.2.6):

$$\nabla \times H = \nabla \times \left(\frac{B}{\mu(B)} \right) = \nabla \times \left(\frac{1}{\mu(B)} \nabla \times A \right) = j \tag{5.2.12}$$

If we consider a linear behaviour of the material, we assume a constant value of magnetic permeability and we can write, using the vectorial identity (5.2.4):

$$\nabla \times (\nabla \times A) = -\nabla^2 A + \nabla(\nabla \cdot A) = \mu \cdot j \tag{5.2.13}$$

Therefore, it is easy to understand that the simplest choice is to assume that:

$$\nabla \cdot A = 0 \tag{5.2.14}$$

which is the so-called "Coulomb gauge". We obtain a Poisson equation:

$$-\nabla^2 A = \mu \cdot j \tag{5.2.15}$$

For understanding better the condition (5.2.14), it is necessary to recall that Helmholtz theorem states also that a vectorial field can be written as the summation of a conservative and a solenoidal field. Let us write the conditions for defining A_1:

$$\begin{cases} \nabla \times A_1 = B \\ \nabla \cdot A_1 = g \end{cases} \tag{5.2.16}$$

with g a scalar field. Now let us define a conservative field:

$$\begin{cases} \nabla \times A_c = 0 \\ \nabla \cdot A_c = g \end{cases} \qquad (5.2.17)$$

Let us define a solenoidal field:

$$\begin{cases} \nabla \times A_S = B \\ \nabla \cdot A_S = 0 \end{cases} \qquad (5.2.18)$$

So we can write:

$$A_1 = A_S + A_C \qquad (5.2.19)$$

because:

$$\begin{cases} \nabla \times A_1 = \nabla \times A_S + \nabla \times A_C = B \\ \nabla \cdot A_1 = \nabla \cdot A_S + \nabla \cdot A_C = g \end{cases} \qquad (5.2.20)$$

Considering the vectorial identity (5.2.2), if we now compare (5.2.8) with (5.2.19) we understand that:

$$\begin{aligned} \nabla \phi &= A_C \\ A &= A_S \end{aligned} \qquad (5.2.21)$$

Therefore, we can write the (5.2.14).

However, for nonlinear materials it is necessary to solve equation (5.2.12). Now we have to define boundary conditions for equation (5.2.12) or (5.2.15). Let us take a look to Figure 5.2-1, which shows a typical representation of a solenoid in order to perform a FEM computation. We notice that there is only the representation of a half section of the solenoid. In fact, we assume an axisymmetric geometry and solution.

We write equation (5.2.7) in cylindrical coordinates.

$$\overline{B} = \hat{r}\left(\frac{1}{r}\frac{\partial A_z}{\partial \phi} - \frac{\partial A_\phi}{\partial z} \right) + \hat{\phi}\left(\frac{\partial A_r}{\partial z} - \frac{\partial A_z}{\partial r} \right) + \hat{z}\frac{1}{r}\left(\frac{\partial (rA_\phi)}{\partial r} - \frac{\partial A_r}{\partial \phi} \right) \qquad (5.2.22)$$

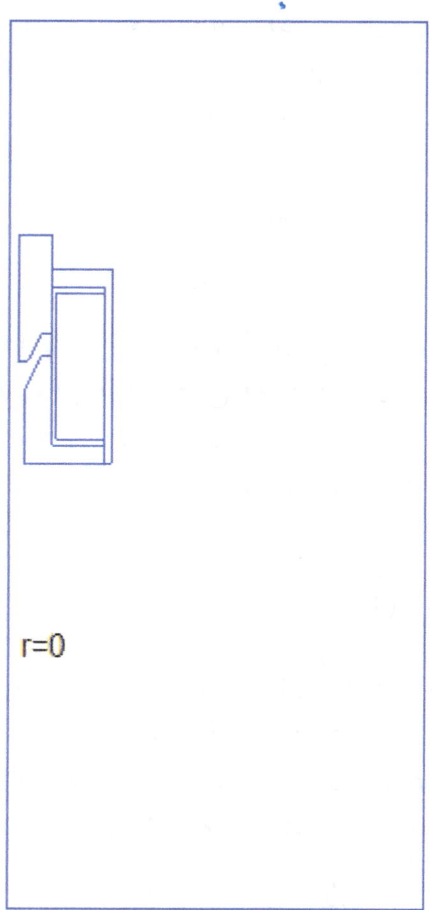

Figure 5.2-1: Solenoid representation.

For an axisymmetric problem, we have that:

$$\frac{\partial}{\partial \phi}(\ ) = 0$$

$$\overline{B} \cdot \hat{\phi} = 0$$

(5.2.23)

From the second equation of (5.2.23) we assume $A_z = A_r = 0$. Therefore, we can write the following expression for B:

$$\overline{B} = \hat{r}\left(-\frac{\partial A_\phi}{\partial z}\right) + \hat{z}\left(\frac{\partial A_\phi}{\partial r} + \frac{A_\phi}{r}\right)$$

(5.2.24)

Therefore, we understand that with these assumptions the vector potential has only one component, which is normal to the plane of solenoid section, while magnetic induction completely lies on this plane. Thanks to equation (5.2.24), we can define the boundary conditions for the vector potential:

$$A_\phi\big|_{\partial D} = 0 \qquad\qquad (5.2.25)$$

In Figure 5.2-1 we can see that the boundary is represented by a rectangle around the solenoid, with one side positioned in r=0. Let us try to understand what this condition means. For the vertical line r=0, if we assume A_φ=0, we have that along this line:

$$\begin{cases} \dfrac{\partial A_\phi}{\partial z} = 0 \\ A_\phi = 0 \end{cases} \rightarrow \overline{B} = \hat{z}\left(\dfrac{\partial A_\phi}{\partial r}\right) \qquad\qquad (5.2.26)$$

Equation (5.2.26) is valid even for the vertical line ideally positioned in r=∞ and it means that magnetic induction has a vertical direction and does not cross the lines. For the horizontal lines above and under the solenoid, if we assume A_φ=0, we can write that:

$$\begin{cases} \dfrac{\partial A_\phi}{\partial r} = 0 \\ A_\phi = 0 \end{cases} \rightarrow \overline{B} = \hat{r}\left(\dfrac{\partial A_\phi}{\partial z}\right) \qquad\qquad (5.2.27)$$

Equation (5.2.27) means that magnetic induction has a horizontal direction and does not cross the boundary lines. These considerations are valid more generally for a boundary of any geometry. So we can easily understand that, assuming A_φ=0 along the boundary means that all the flux lines close inside the solution domain (no line crosses the boundary). This assumption means that it is necessary to position the upper, lower and right lines far from the solenoid.

Now we want to write in the explicit form the equation (5.2.15). This is a vectorial equation because A and j are vectors. We demonstrate that in the case of an axisymmetric problem we have only one equation. First of all, we write the expression of the Vector Laplacian:

$$\nabla^2 A = \left(\Delta A_r - \frac{A_r}{r^2} - \frac{2}{r^2}\frac{\partial A_\phi}{\partial \phi} \right)\hat{r} + \left(\Delta A_\phi - \frac{A_\phi}{r^2} + \frac{2}{r^2}\frac{\partial A_r}{\partial \phi} \right)\hat{\phi} + \Delta A_z \hat{z}$$

(5.2.28)

where ΔA is the Laplace operator. Considering the fact that the solution is axisymmetric, we can write:

$$\nabla^2 A = \left(\Delta A_\phi - \frac{A_\phi}{r^2} \right)\hat{\phi}$$

(5.2.29)

Therefore, it is easy to understand that, since the current vector j has the only component in the ϕ direction, we have to solve only one equation. Now we write the Laplace operator for A_ϕ, which will be called directly with the symbol of A, for an axisymmetric solution:

$$\Delta A_\phi = \frac{1}{r}\frac{\partial}{\partial r}\left(r\frac{\partial A}{\partial r} \right) + \frac{1}{r^2}\frac{\partial^2 A}{\partial \phi^2} + \frac{\partial^2 A}{\partial z^2} =$$
$$= \frac{1}{r}\left(r\frac{\partial^2 A}{\partial r^2} + \frac{\partial A}{\partial r} \right) + \frac{\partial^2 A}{\partial z^2} = \frac{\partial^2 A}{\partial r^2} + \frac{1}{r}\frac{\partial A}{\partial r} + \frac{\partial^2 A}{\partial z^2}$$

(5.2.30)

Therefore, we can write the scalar equation (5.2.15) in the following way:

$$-\left(\frac{\partial^2 A}{\partial r^2} + \frac{1}{r}\frac{\partial A}{\partial r} + \frac{\partial^2 A}{\partial z^2} - \frac{A}{r^2} \right) = \mu \cdot j$$

(5.2.31)

5.3. FEM Formulation

Finite element method is based on two different equivalent formulations. Both of them are based on solving the governing equations of the problem in the integral form. For understanding better what we are talking about, we will give a short description of both formulation.

First Formulation: weak solutions

A general unidimensional Poisson problem can be represented in the following way:

$$\begin{cases} -u_{xx} = f(x) \\ u(0) = 0 \end{cases}, \ x \in [0,1] \tag{5.3.1}$$

A classical solution of this problem is a function u that, for $x \in [0,1]$, is such that $u \in C_2$. It means that u, its first and second derivatives are continuous for $x \in [0,1]$. As we can see from (5.3.1), this is possible only if the function $f(x)$ is continuous in the interval. This condition, for problems describing physical system is rare to be satisfied. Poisson problems describe usually diffusion problems, such as thermal conduction in solids. In such a case, $f(x)$ represents a heat generation term that is usually confined in a restricted area and so it cannot be continuous in the dominium. If we analyse the magnetostatic problem, we have to look at equation (5.2.15): $f(x)$ is the current density that is present only in the region where the coils are present.

Therefore, we can easily understand that we have to use a different approach. This approach consists in looking for a solution that does not have to be regular even for the second derivative. The complete mathematical description of the method we want to use is too much complicated to explain because it involves the knowledge of concepts like Lebesgue integral and Sobolev spaces. On the contrary, we will just describe the basic concepts useful in order to better understand the mathematical background of the FEM method.

The approach consists in looking for a solution such that $u \in C_1$ for $x \in [0,1]$. In the case of equation (5.2.15), it means that we will find a continuous magnetic induction field. First of all, we multiply both terms of equation (5.3.1) for an arbitrary function $v \in C_\infty$ for $x \in [0,1]$ and integrate the results in the domain of the problem:

$$-\int_0^1 u_{xx} v(x)dx = \int_0^1 f(x)v(x)dx \qquad (5.3.2)$$

This function is called "test function" which has been introduced just for one reason: we will find the function $u(x)$ which solves the integral equation in (5.3.2). We do this operation because this new problem does not need that the second derivative of u is continuous. For this reason, the solution of the problem is called "weak solution" because we have decreased the order of the problem. We apply the Green Theorem:

$$\int_0^1 u_{xx} v\,dx = \int_0^1 u_x v_x - [u_x v]_0^1 \qquad (5.3.3)$$

We arbitrary choose $v(x)$ such that $v(0)=0$ and $v(1)=0$. So we can write:

$$-\int_0^1 u_x v_x dx = \int_0^1 f(x)v(x)dx \qquad (5.3.4)$$

Now it is clear that we have decreased the order of the problem because only first derivatives appear in the problem. Now we simplify the notations:

$$\begin{cases} a(u,v) = -\int_0^1 u_x v_x dx \\ (f,v) = \int_0^1 f(x)v(x)dx \\ a(u,v) = (f,v) \end{cases} \qquad (5.3.5)$$

In order to solve (5.3.5) we introduce the method of Galerkin. It consists in solving in a discrete and approximated form the equation (5.3.5). First of all, we write the approximated solution as:

$$u_h = \sum_{i=1}^N \xi_i \varphi_i(x) \qquad (5.3.6)$$

where φ_i for $i=1,\ldots,N$ (N is the order of discretization) are functions linearly independent (a basis in the space of solution). These functions are typically polynomials of the first order (linear Galerkin method) or of the second order (nonlinear Galerkin method). They can be defined in the following way:

$$\varphi_i(x) = \begin{cases} \dfrac{x - x_{i-1}}{x_i - x_{i-1}}, & x_{i-1} \le x \le x_i \\[2mm] \dfrac{x_{i+1} - x}{x_{i+1} - x_i}, & x_i \le x \le x_{i+1} \\[2mm] 0, & elsewhere \end{cases} \tag{5.3.7}$$

Represented in Figure 5.3-2, they bring to a linear approximation described by Figure 5.3-1:

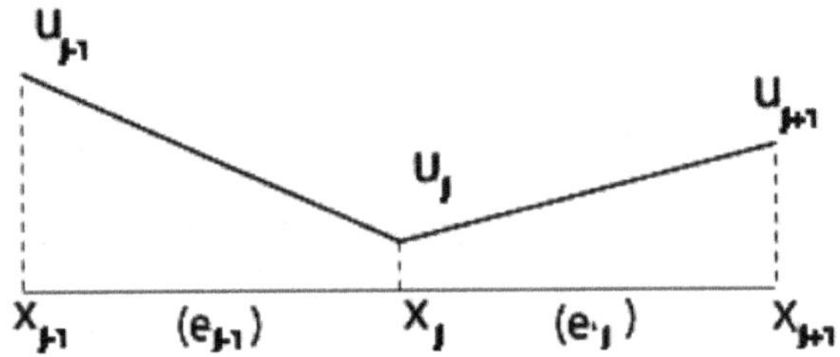

Figure 5.3-1: Linear discrete approximation of the solution (from [14]).

The next step is to choose as arbitrary function *v(x)* the functions φᵢ(x):

$$v_i = \varphi_i(x) \tag{5.3.8}$$

Therefore, we automatically satisfy:

$$\begin{cases} \varphi_1(0) = 0 \\ \varphi_N(1) = 0 \end{cases} \tag{5.3.9}$$

Now we solve the following equations:

$$a(u_h, \varphi_i) = (f, \varphi_i) \; ; \; i = 1, ..., N \tag{5.3.10}$$

If we explicit the equation (5.3.10), we find the following linear algebraic system:

$$\sum_{j=1}^{N} \alpha_{ij} \xi_j = \beta_i \; ; \; i = 1, ..., N$$

$$\alpha_{ij} = a(\varphi_i, \varphi_j) \tag{5.3.11}$$

$$\beta_i = (f, \varphi_i)$$

We have to find the unknown quantity ξ_i.

The method explained above is for one-dimensional problem, but it gives the possibility to show all the relevant aspects in solving partial differential equation in the weak form. We want to underline that this solution is fully representative of the problem (5.3.1). In fact if we remember how we have found the differential formulation of electro-magnetism (Maxwell equations), we realize that we started from an integral form. Then we have superimposed some mathematical conditions in order to write equations in every point of the space. Therefore, it is easy to understand that the weak solution represents the solution of the integral form of the problem, which is the physical one.

Second formulation: variational approach

The second formulation is based on the theory of variations. In the following, we describe the mathematical formulation of this problem.

For a general Poisson boundary value problem as in (5.3.1):

$$\begin{cases} -\nabla^2 u = f \ \ in \ \ D \\ u = g \ \ \ in \ \ \partial D \end{cases} \qquad (5.3.12)$$

We define the energy functional F:

$$F(u) = \int_D \left[\frac{1}{2} |\nabla u|^2 - f \cdot u \right] dxdydz \qquad (5.3.13)$$

where:

$$|\nabla u|^2 = \left(\frac{\partial u}{\partial x} \right)^2 + \left(\frac{\partial u}{\partial y} \right)^2 + \left(\frac{\partial u}{\partial y} \right)^2 \qquad (5.3.14)$$

It can be demonstrated that the function u_{SOL} such as:

$$F(u_{SOL}) = \min F(u) \qquad (5.3.15)$$

is the solution of (5.3.12). Moreover, it can be demonstrated that if a boundary value problem admit a functional, the solution of the minimization problem is the same solution of the weak formulation. Therefore, we can affirm that the Galerkin method and the variational method converge to the solution of the problem.

Now we apply these equations to a 1D problem such as (5.3.1):

$$F(u) = \int_D \left[\frac{1}{2}\left(\frac{\partial u}{\partial x}\right)^2 - f \cdot u \right] dx \tag{5.3.16}$$

We write the approximated solution as in (5.3.6). Then we apply the rules of variational calculus, which states that the function u that minimize the functional is such as the first variation of F is zero. So we write:

$$\frac{\partial F}{\partial \xi_i} = \frac{\partial}{\partial \xi_i} \left[\int_0^1 \frac{1}{2}\left(\sum_{i=1}^{N} \xi_i \frac{\partial \varphi_i}{\partial x}\right)^2 dx - \int_0^1 f \cdot \left(\sum_{i=1}^{N} \xi_i \varphi_i\right) dx \right] = 0 \; ; \; i = 1,....,N \tag{5.3.17}$$

So we can write:

$$\int_0^1 \frac{\partial \varphi_i}{\partial x}\left(\sum_{j=1}^{N} \xi_j \frac{\partial \varphi_j}{\partial x}\right) dx = \int_0^1 f \cdot \varphi_i dx \; ; \; i = 1,....,N \tag{5.3.18}$$

which is linear system $AX=B$ with the following coefficients:

$$a_{ij} = \int_0^1 \frac{\partial \varphi_i}{\partial x}\frac{\partial \varphi_j}{\partial x} dx$$

$$b_i = \int_0^1 f \cdot \varphi_i dx \tag{5.3.19}$$

$$X = \left[\xi_i,...,\xi_N\right]^T$$

Choosing the continuous piecewise linear approximation of (5.3.7), means that we are calculating the value of the solution in the nodes of the discretization (see Figure 5.3-2):

$$u_h = \sum_{i=1}^{N} u_i \varphi_i(x) \tag{5.3.20}$$

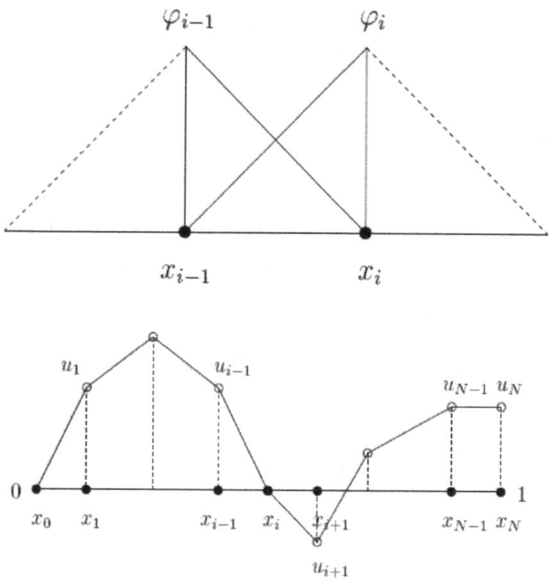

Figure 5.3-2: Piecewise linear approximation (from [15]).

If we choose a uniform mesh $\Delta x = \dfrac{1}{N}$ and $f=1$:

$$A = \frac{1}{(\Delta x)^2}\begin{bmatrix} 2 & -1 & & & \\ -1 & 2 & -1 & & \\ & \cdot & \cdot & \cdot & \\ & & -1 & 2 & -1 \\ & & & -1 & 1 \end{bmatrix} \; ; \; B = \begin{bmatrix} 1 \\ 1 \\ \cdot \\ 1 \\ 1/2 \end{bmatrix} \; ; \; X = \begin{bmatrix} u_1 \\ u_2 \\ \cdot \\ u_{N-1} \\ u_N \end{bmatrix}$$

$$(5.3.21)$$

Now let us focus on a 2D problem in order to understand discretization of the geometry. We focus on a pure 2D problem but the concepts will be the same even for a general 3D axis symmetrical problem like the one described by equation (5.2.31). The problem is to solve a 2D magnetostatic field, whose functional is:

$$F(A) = \int_D \left[\frac{1}{2\mu} |\nabla A|^2 - J \cdot A \right] dx dy \qquad (5.3.22)$$

where A is the magnetic potential (in a 2D case has only the component normal to the plane), μ is the magnetic permeability (which is constant because we consider a linear problem) and J is the current density (which represents for example a coil). For this kind of problem, we discretize the geometry with a typical triangle mesh (Delaunay triangulation) and so for each element we can write a linear approximation as:

$$A^{(z)} = \sum_{k=L,M,N} A_k \left(a_k + b_k x + c_k y \right) \qquad (5.3.23)$$

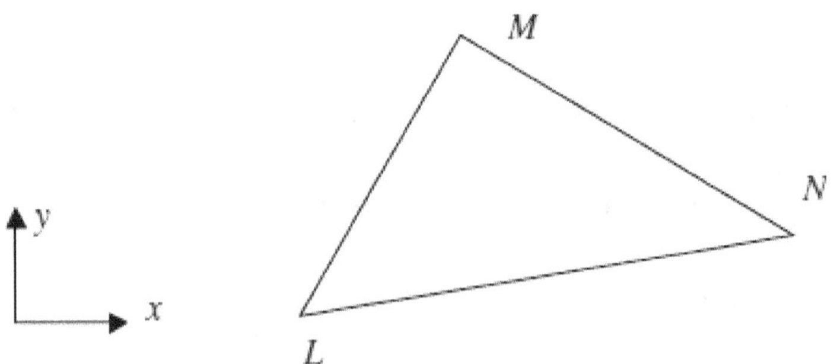

Figure 5.3-3: Triangle element (from [12]).

A_k is the value of the potential in the nodes of the element and the letter "z" identify the element (see Figure 5.3-3). We find the expression for the coefficient a_k, b_k, c_k. We know the coordinates of the nodes of the element and so we can impose the value of the potential on the nodes. We find:

$$\begin{pmatrix} a_L & a_M & a_N \\ b_L & b_M & b_N \\ c_L & c_M & c_N \end{pmatrix} = \begin{pmatrix} 1 & x_L & y_L \\ 1 & x_M & y_M \\ 1 & x_N & y_N \end{pmatrix}^{-1} \qquad (5.3.24)$$

Therefore, it is easy to understand that for every element there is a different approximation. The expression (5.3.23) represents the corresponding 2D approximation of the 1D one represented in Figure 5.3-2. In fact, we can represent this approximation as in Figure 5.3-4 where is shown the generic function $\varphi_L = A_L(a_L + b_L x + c_L y)$.

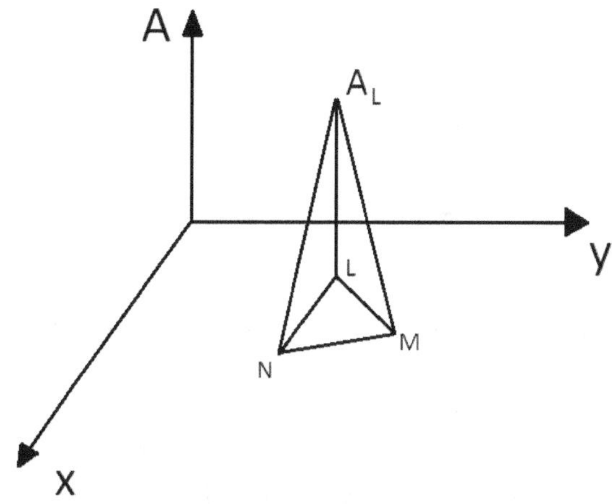

Figure 5.3-4: 2D linear approximation.

Now we write the functional only for the generic element "z":

$$F^{(z)}(A) = \int_{D^z} \left\{ \frac{1}{2\mu^{(z)}} \left[\left(\frac{\partial A^{(z)}}{\partial x} \right)^2 + \left(\frac{\partial A^{(z)}}{\partial y} \right)^2 \right] - J^{(z)} \cdot A^{(z)} \right\} dx dy$$

$$(5.3.25)$$

where $J(z)$ is the current density in the element. Now we write the derivatives of $A^{(z)}$:

$$\frac{\partial A^{(z)}}{\partial x} = \sum_k A_k b_k \quad ; \quad \frac{\partial A^{(z)}}{\partial y} = \sum_k A_k c_k \qquad (5.3.26)$$

So we can write:

$$\left(\frac{\partial A^{(z)}}{\partial x} \right)^2 = \left(A_L b_L + A_M b_M + A_N b_N \right)^2 =$$

$$= A_L^2 b_L^2 + A_M^2 b_M^2 + A_N^2 b_N^2 + 2 A_L b_L A_M b_M + 2 A_L b_L A_N b_N + 2 A_M b_M A_N b_N = C_X$$

$$(5.3.27)$$

$$\left(\frac{\partial A^{(z)}}{\partial y}\right)^2 = \left(A_L c_L + A_M c_M + A_N c_N\right)^2 =$$

$$= A_L^2 c_L^2 + A_M^2 c_M^2 + A_N^2 c_N^2 + 2 A_L c_L A_M c_M + 2 A_L c_L A_N c_N + 2 A_M c_M A_N c_N = C_Y$$

$$(5.3.28)$$

Substituting (5.3.27) and (5.3.28):

$$F^{(z)}(A) = \int\limits_{D^z}\left[\frac{1}{2\mu^{(z)}}\left(C_X + C_Y\right) - J^{(z)} \cdot A^{(z)}\right]dxdy =$$

$$\frac{S^{(z)}}{2\mu^{(z)}}\left(C_X + C_Y\right) - J^{(z)} \cdot \int\limits_{D^z}\left[\sum_k A_k\left(a_k + b_k x + c_k y\right)\right]dxdy$$

$$(5.3.29)$$

where $S^{(z)}$ is the surface area of the element. Now we write the integral in (5.3.29):

$$\int\limits_{D^z}\left[\sum_k A_k\left(a_k + b_k x + c_k y\right)\right]dxdy = \sum_k \int\limits_{D^z}\left[A_k\left(a_k + b_k x + c_k y\right)\right]dxdy$$

$$(5.3.30)$$

As we have shown in Figure 5.3-4, the generic function $\varphi_K = A_K(a_K + b_K x + c_K y)$ describes the surface of a pyramid. Therefore, it is easy to understand that the integral in (5.3.30) represents the volume of the pyramid:

$$\int\limits_{D^z}\left[A_k\left(a_k + b_k x + c_k y\right)\right]dxdy = \frac{S^{(z)}}{3} A_k \qquad (5.3.31)$$

So we can write:

$$F^{(z)}(A) = \frac{S^{(z)}}{2\mu^{(z)}}\left(C_X + C_Y\right) - J^{(z)} \cdot \frac{S^{(z)}}{3}\sum_k A_k \qquad (5.3.32)$$

Now we can minimize the functional of the element in order to solve the magnetostatic problem. We write the following equations:

$$\begin{cases} \dfrac{\partial F^{(z)}}{\partial A_L} = 0 \\[2mm] \dfrac{\partial F^{(z)}}{\partial A_M} = 0 \\[2mm] \dfrac{\partial F^{(z)}}{\partial A_N} = 0 \end{cases} \tag{5.3.33}$$

Solving (5.3.33), we obtain the following 3x3 system of equations:

$$\frac{S^{(z)}}{\mu^{(z)}} \begin{bmatrix} b_L^2 + c_L^2 & b_L b_M + c_L c_M & b_L b_N + c_L c_N \\ b_L b_M + c_L c_M & b_M^2 + c_M^2 & b_M b_N + c_M c_N \\ b_L b_N + c_L c_N & b_M b_N + c_M c_N & b_N^2 + c_N^2 \end{bmatrix} \begin{Bmatrix} A_L \\ A_M \\ A_N \end{Bmatrix} = \frac{S^{(z)}}{3} J^{(z)} \begin{Bmatrix} 1 \\ 1 \\ 1 \end{Bmatrix} \tag{5.3.34}$$

This system represents the elementary equations for the solution of the problem. We can recognize an array of unknown variables q_z, a stiffness matrix k_z and an array of input F_z.

$$k_z q_z = F_z \tag{5.3.35}$$

Now we have to solve the problem for the entire geometry. Therefore, we have to write all the elementary system of equations for all the geometry. The problem is that, in a triangular element mesh, every node is shared by 3 or 4 triangles. Therefore, it is easy to understand that we cannot solve the elementary system separately but all together. It means that we have to build a general expression for (5.3.35). We can define the following entities:

$$K_d = \begin{bmatrix} [k_1] & 0 & 0 & \dots \\ 0 & [k_2] & 0 & \dots \\ 0 & 0 & [k_3] & \dots \\ \dots & 0 & 0 & \dots \end{bmatrix} ; \; Q_d = \begin{bmatrix} [q_1] \\ [q_2] \\ [q_3] \\ \dots \end{bmatrix} ; \; F_d = \begin{bmatrix} [F_1] \\ [F_2] \\ [F_3] \\ \dots \end{bmatrix} \tag{5.3.36}$$

Therefore, we can write the general solution as:

$$K_d Q_d = F_d \tag{5.3.37}$$

Now we have to define what is called "Connectivity matrix". We define the array of the solution as:

$$Q = \begin{bmatrix} A_1 \\ A_2 \\ A_3 \\ \cdots \end{bmatrix} \tag{5.3.38}$$

Now we look at the Figure 5.3-5, which represents a general simple geometry:

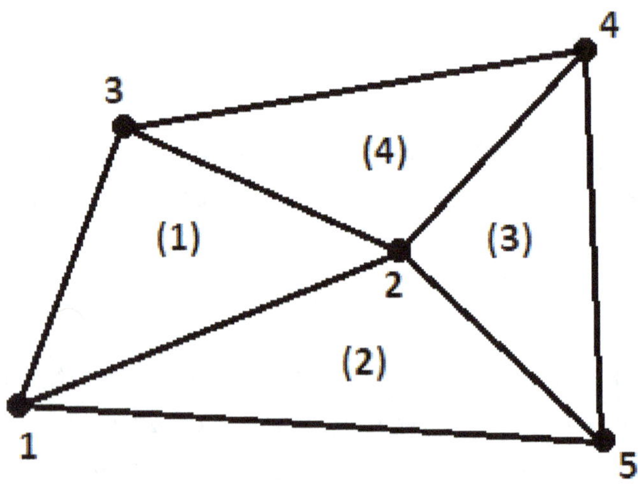

Figure 5.3-5: Simple mesh geometry.

For this geometry, we want to find the "Connectivity matrix" "A" such as:

$$Q_d = A \cdot Q \tag{5.3.39}$$

It is easy to write this matrix for Figure 5.3-5:

$$Q_d = \begin{Bmatrix} q_1 \\ q_2 \\ q_3 \\ q_1 \\ q_2 \\ q_5 \\ q_2 \\ q_4 \\ q_5 \\ q_2 \\ q_3 \\ q_4 \end{Bmatrix} = \begin{bmatrix} 1 & 0 & 0 & 0 & 0 \\ 0 & 1 & 0 & 0 & 0 \\ 0 & 0 & 1 & 0 & 0 \\ 1 & 0 & 0 & 0 & 0 \\ 0 & 1 & 0 & 0 & 0 \\ 0 & 0 & 0 & 0 & 1 \\ 0 & 1 & 0 & 0 & 0 \\ 0 & 0 & 0 & 1 & 0 \\ 0 & 0 & 0 & 0 & 1 \\ 0 & 1 & 0 & 0 & 0 \\ 0 & 0 & 1 & 0 & 0 \\ 0 & 0 & 0 & 1 & 0 \end{bmatrix} \begin{Bmatrix} q_1 \\ q_2 \\ q_3 \\ q_4 \\ q_5 \end{Bmatrix} \qquad (5.3.40)$$

Now we elaborate the equation (5.3.37) using (5.3.39):

$$K_d Q_d = F_d \rightarrow K_d A \cdot Q = F_d \rightarrow A^T K_d A \cdot Q = A^T F_d \qquad (5.3.41)$$

Therefore, we can write the problem as:

$$K \cdot Q = F$$
$$K = A^T K_d A \qquad (5.3.42)$$
$$F = A^T F_d$$

which represents the general system of equations solving the magnetostatic field of the entire geometry. We want to underline that all the FEM software follow a different assembly procedure, which is faster than the one discussed herein. Moreover, for cylindrical solenoid, we have to solve an axial-symmetry problem which has a formulation as per (5.2.31), and it requires a different approach.

In the following figures, a typical solenoid mesh is shown in order to appreciate the general FEM formulation and element distribution.

We can notice the higher mesh density in air gaps. That's because in this regions magnetic permeability is discontinuous: its values in the ferromagnetic material is typically about 1000 ÷ 8000 while in air it has an unitary value. So we can understand that the solution gradients are very high in this region and so it is necessary to have a higher mesh density, typically a minimum of three layer are necessary.

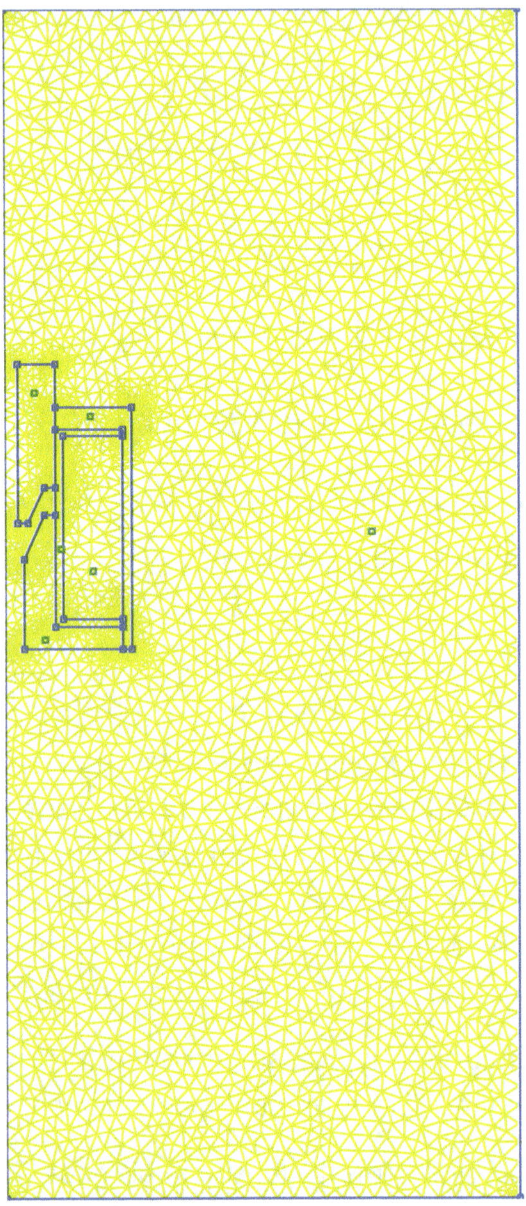

Figure 5.3-6: Solenoid mesh geometry.

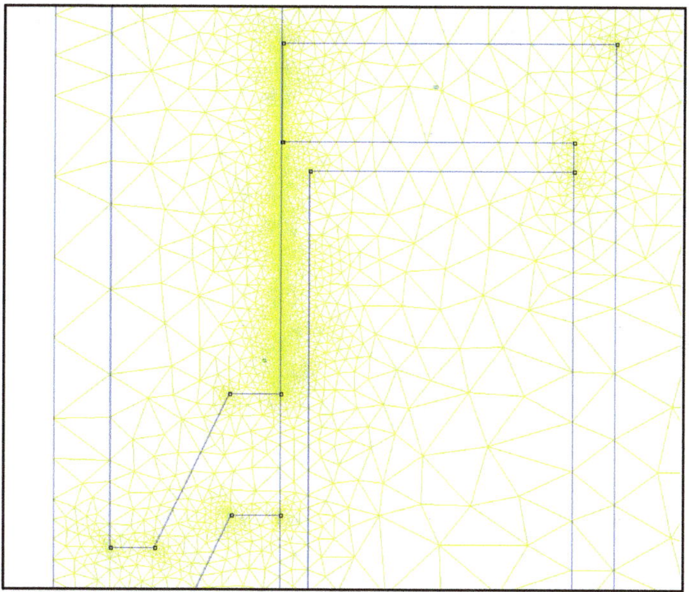

Figure 5.3-7: Solenoid mesh geometry: detail.

Chapter 6

Thermal analysis

6.1. Introduction

In solenoid design, it is very important to evaluate thermal performances of the solenoid. These performances influence operative cycles of solenoid. In fact, if a solenoid has a good heat exchange and it never heats up to dangerous temperatures, it can be used for an indefinite time. Therefore, this kind of solenoid is called a "Continuous duty" solenoid. The parameter that shall be taken in account is the coil temperature, which must be less than thermal class temperature (see [1]). If a solenoid heats up to dangerous temperature after some time, it is called "intermittent duty" solenoid: an operating cycle is defined which is a balance of warming when it is energized and cooling when it is switched off.

We will define two models to compute thermal performances of solenoid, a stationary one and a transitory one. The second model is an elaboration of the first one.

6.2. Stationary thermal model

The goal of this model is to compute the stationary temperature of the solenoid coil when energized for an indefinite time. This kind of computation gives the possibility to understand if the solenoid is able to operate in a "continuous duty". If the temperature is too high, it is necessary to understand time dependency of coil temperature to establish the solenoid operating cycle. In that case, the second model will be introduced.

First of all, we examine the geometry of a typical solenoid in Figure 6.2-1.

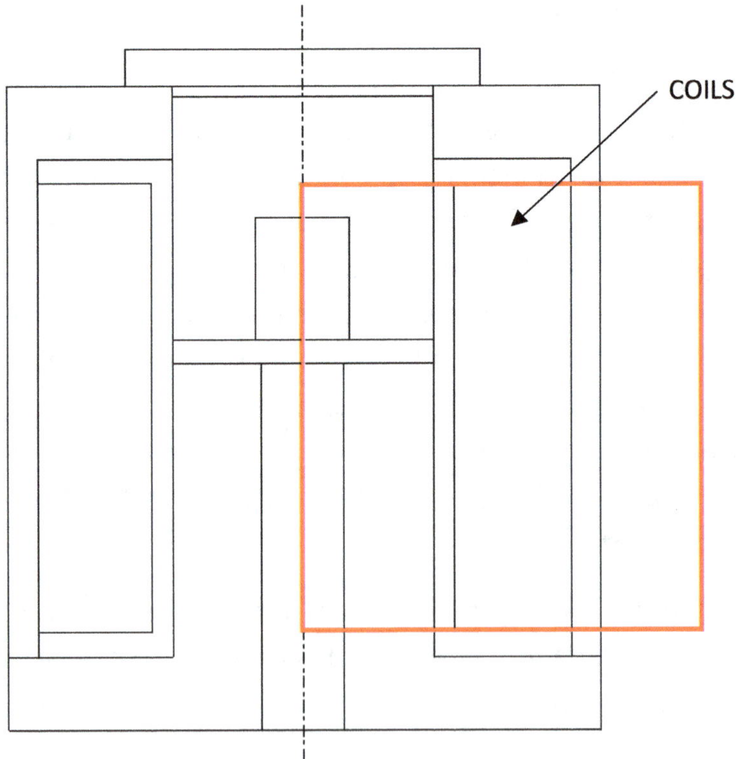

Figure 6.2-1: Solenoid geometry.

We make the following assumption:

- Heat exchange is possible only in radial direction. It means that we consider that on the top and on the bottom of the solenoid, there is an infinite thermal resistance. This is not true but the higher thickness of material in axial direction respect to radial direction gives rise to a higher thermal resistance. Moreover, on the top and on the bottom of the solenoid an insulant (from electrical and thermal point of view) fabric is installed, giving rise to a higher thermal resistance (see Figure 6.2-2).
- The temperature of the coil is uniform from windings to the axis.

Therefore, we can approximate thermal field as an axis symmetric cylindrical field. Therefore, we are interested only in a part of the geometry that has been highlighted in red in Figure 6.2-1. This box represents the part of the total geometry we are interested. In Figure 6.2-2, we can see a zoom of the zone.

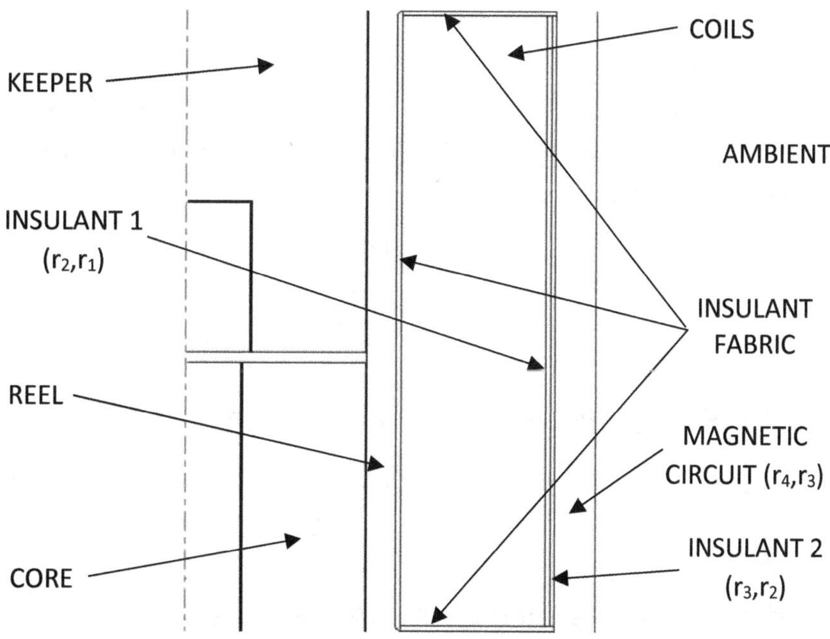

Figure 6.2-2: Detail of the geometry.

Therefore, we have to solve the Fourier equation in cylindrical coordinates:

$$\frac{d^2T}{dr^2} + \frac{1}{r}\frac{dT}{dr} = 0 \tag{6.2.1}$$

If we integrate (6.2.1) and explicit thermal power exchanged in radial direction for a cylindrical pipe, we obtain:

$$W = \lambda 2\pi Y \frac{\left(T_2 - T_1\right)}{\ln\left(\dfrac{r_2}{r_1}\right)} \tag{6.2.2}$$

where:

- λ is thermal conductivity (W/(m K));
- r_2 is the outer radius of the pipe;
- r_1 is the inner radius of the pipe;
- Y is the height of windings (see chapter 2);
- T_2 is the outer temperature;
- T_1 is the inner temperature.

Heat flows from coils to external ambient so we can write (6.2.2) for every layer. The first layer is represented by the insulant 1, typically a fabric placed around coils or a sealant:

$$W = \lambda_1 2\pi Y \frac{\left(T_1 - T_2\right)}{\ln\left(\dfrac{r_2}{r_1}\right)} \tag{6.2.3}$$

The second layer is represented by the insulant 2, typically a sealant:

$$W = \lambda_2 2\pi Y \frac{\left(T_3 - T_2\right)}{\ln\left(\dfrac{r_3}{r_2}\right)} \tag{6.2.4}$$

The third layer is represented by the magnetic case:

$$W = \lambda_3 2\pi Y \frac{(T_4 - T_3)}{\ln\left(\dfrac{r_4}{r_3}\right)} \tag{6.2.5}$$

Finally, thermal power is dissipated with external ambient by natural convection and radiation. If we name K the total heat exchange coefficient with external ambient (W/(m²K)) and T_{amb} the ambient temperature, we can write:

$$W = K 2\pi r_4 Y (T_4 - T_{amb}) \tag{6.2.6}$$

From (6.2.3) to (6.2.6) we can rearrange the equations and explicit the internal temperature T_1 that is equal to the coils temperature T_{COILS}:

$$T_{COILS} = T_{AMB} + W \cdot H$$

$$H = \left(\frac{\ln\left(\dfrac{r_2}{r_1}\right)}{\lambda_1 2\pi Y} + \frac{\ln\left(\dfrac{r_3}{r_2}\right)}{\lambda_2 2\pi Y} + \frac{\ln\left(\dfrac{r_4}{r_3}\right)}{\lambda_3 2\pi Y} + \frac{1}{K 2\pi Y r_4} \right) \tag{6.2.7}$$

Notice that H is the thermal resistance (K/W). This equation must be coupled with the following equation:

$$W = \frac{V^2}{R_0 \left[1 + \alpha (T_{COILS} - T_0)\right]} \tag{6.2.8}$$

where T_0 is the reference temperature (20°C typically), R_0 electrical resistance at T_0 and V the tension applied.

When we couple (6.2.7) with (6.2.8) we obtain a quadratic equation in T_{COILS}:

$$aT^2 + bT + c = 0$$

$$a = \alpha$$

$$b = 1 - \alpha T_0 - \alpha T_{AMB} \tag{6.2.9}$$

$$c = -T_{AMB} + \alpha T_0 T_{AMB} - \frac{V^2}{R_0} H$$

The solution of (6.2.9) gives the final stationary temperature of the coil.

6.3. Transitory model

For "intermittent duty" solenoids, it is necessary to compute coils temperature as a function of time. We have to elaborate the first model to take into account the effect of thermal capacitances and not only thermal resistances. We follow the same assumptions made in the first model. It means that we consider in every moment a uniform temperature from outer diameter of coils and solenoid axis. Therefore, we have to associate to this part of the solenoid a total mass M and an equivalent specific heat C (computed in §6.6).

Therefore, we write heat equation as:

$$CM\frac{dT}{dt} = \frac{V^2}{R_0[1+\alpha(T-T_0)]} - \frac{(T-T_{AMB})}{H} \qquad (6.3.1)$$

We rearrange the equation:

$$CM\frac{dT}{dt} = \frac{V^2 - \dfrac{R_0}{H}[1+\alpha(T-T_0)](T-T_{AMB})}{R_0[1+\alpha(T-T_0)]} \qquad (6.3.2)$$

$$\frac{dt}{CMR_0} = \frac{1+\alpha(T-T_0)}{V^2 - \dfrac{R_0}{H}[1+\alpha(T-T_0)](T-T_{AMB})}dT \qquad (6.3.3)$$

$$\frac{dt}{CMR_0} = \frac{1+\alpha(T-T_0)}{V^2 - \dfrac{R_0}{H}\left[(\alpha)T^2 + (1-\alpha T_0 - \alpha T_{AMB})T + (\alpha T_0 T_{AMB} - T_{AMB})\right]}dT$$

$$(6.3.4)$$

Now that we have separated the variables, we are able to integrate temperature from T_{AMB} and T and time from 0 to t.

We finally write:

$$\int_{T_{AMB}}^{T} \frac{d+eT}{aT^2+bT+c}\,dT = \int_{0}^{t} \frac{dt}{CMR_0}$$

$$a=-\frac{R_0}{H}\alpha$$

$$b=-\frac{R_0}{H}\left(1-\alpha T_0 - \alpha T_{AMB}\right) \tag{6.3.5}$$

$$c=V^2-\frac{R_0}{H}\left(\alpha T_0 T_{AMB} - T_{AMB}\right)$$

$$d=1-\alpha T_0$$

$$e=\alpha$$

We compute the primitive function of the first member of (6.3.5). We define the discriminant as

$$\Delta = b^2 - 4ac \tag{6.3.6}$$

It is possible to demonstrate that:

$$if \begin{cases} \alpha = 0.0039 K^{-1} \\ T_0 = 293.15K \end{cases} \rightarrow \Delta \geq 0 \tag{6.3.7}$$

There is only a possibility that $\Delta=0$:

$$if \begin{cases} V=0 \\ T_{AMB} = T_0 - \dfrac{1}{\alpha} = 36.73K \end{cases} \rightarrow \Delta = 0 \tag{6.3.8}$$

It is easy to understand that the conditions in (6.3.8) are without interest. So if $\Delta>0$, there exist real and distinct solutions for the equation $aT^2+bT+c=0$. The solutions are:

$$T_1 = \frac{-b-\sqrt{\Delta}}{2a}$$

$$T_2 = \frac{-b+\sqrt{\Delta}}{2a} \tag{6.3.9}$$

Now we apply the method of partial fraction decomposition in order to solve the integral. We can write:

$$\frac{d+eT}{aT^2+bT+c}=\frac{\frac{d}{a}+\frac{e}{a}T}{(T-T_1)(T-T_2)} \tag{6.3.10}$$

We can write (6.3.10) in this way:

$$\frac{d+eT}{aT^2+bT+c}=\frac{\frac{d}{a}+\frac{e}{a}T}{(T-T_1)(T-T_2)}=\frac{A}{(T-T_1)}+\frac{B}{(T-T_2)}$$

$$A=-\frac{\frac{d}{a}+\frac{e}{a}T_1}{(T_2-T_1)} \tag{6.3.11}$$

$$B=\frac{\frac{d}{a}+\frac{e}{a}T_2}{(T_2-T_1)}$$

Now we can write the solution of the integral at the first member of (6.3.5):

$$\int\frac{d+eT}{aT^2+bT+c}dT=A\ln(T-T_1)+B\ln(T-T_2)+COSTANT \tag{6.3.12}$$

With this solution, (6.3.5) can be written as:

$$\ln\left[\left(\frac{T-T_1}{T_{AMB}-T_1}\right)^A\left(\frac{T-T_2}{T_{AMB}-T_2}\right)^B\right]=\frac{t}{CMR_0} \tag{6.3.13}$$

It is easy to understand that explicating coil temperature T from (6.3.13) is impossible. So we can use a different approach: we compute non the function $T=T(t)$ but the inverse function $t=t(T)$:

$$t=CMR_0\ln\left[\left(\frac{T-T_1}{T_{AMB}-T_1}\right)^A\left(\frac{T-T_2}{T_{AMB}-T_2}\right)^B\right] \tag{6.3.14}$$

In Figure 6.3-1, we can see the plot of this function.

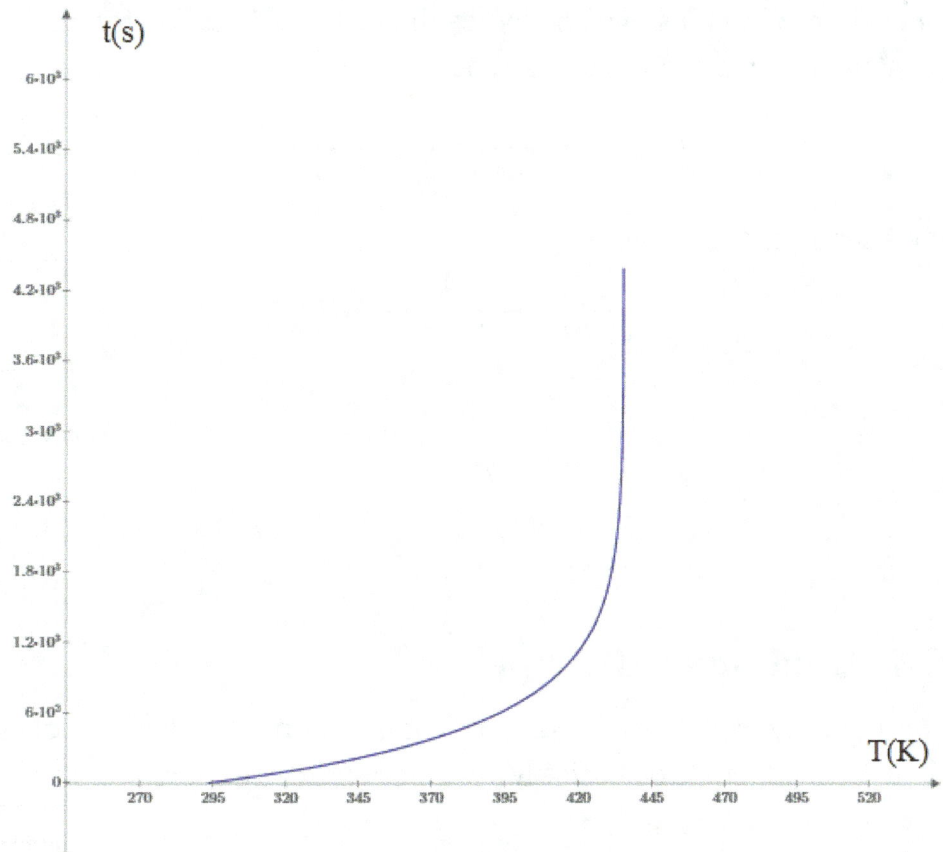

Figure 6.3-1: Time Vs. Coil temperature.

It can be demonstrated that the maximum temperature is T_l. This temperature is the same that we can compute solving the system (6.2.9) in the stationary model, because the quadratic equation in (6.2.9) have the same coefficient of (6.3.5), except from a multiplicative term.

We end the paragraph writing again the system (6.3.5) for a multiple coil solenoid. We write the total initial power "produced" by the solenoid as:

$$W_0 = \sum_i \frac{V_i^2}{R_i} \qquad (6.3.15)$$

where V_i are the tensions applied to the coil "i", while R_i is the coil "i" resistance. So we can write (6.3.1) as:

$$CM\frac{dT}{dt} = \frac{W_0}{1+\alpha(T-T_0)} + \frac{T-T_{AMB}}{H}$$

(6.3.16)

If we separate the variables:

$$\frac{dt}{CMH} = \frac{d+eT}{aT^2+bT+c}dT$$

$$a = -\alpha$$

$$b = -1+\alpha(T_0+T_{AMB})$$

$$c = W_0H+T_{AMB}-\alpha T_0 T_{AMB}$$

$$d = 1-\alpha T_0$$

$$e = \alpha$$

(6.3.17)

6.4. Cooling of the coil

For intermittent duty cycle solenoids, it is important to know cooling time when energized. Basically it is important to know the law of cooling $T=T(t)$.

We have to solve the equation (6.3.1) with the assumption of $V=0$. So we can write:

$$CM\frac{dT}{dt} = -\frac{(T-T_{AMB})}{H}$$

(6.4.1)

Rearranging and integrating the equation between initial temperature T_{in} and temperature at time t we have:

$$\int_{T_{in}}^{T}\frac{dT}{T-T_{AMB}} = -\int_{0}^{t}\frac{dt}{CMH}$$

(6.4.2)

The solution is:

$$\ln\frac{T-T_{AMB}}{T_{in}-T_{AMB}} = -\frac{t}{CMH}$$

(6.4.3)

If we explicit temperature we obtain the well-known Newton law of cooling:

$$T = T_{AMB} + \left(T_{in} - T_{AMB}\right) \cdot e^{-\frac{t}{CMH}}$$
(6.4.4)

6.5. Thermal properties of materials

In order to apply the models we have built in this chapter, we need to know the thermal and physical properties of common materials used in solenoids. The materials we have selected do not cover all the possible choices. However, they are representative to understand the order of magnitude of properties values.

Thermal conductivity

In the following table, thermal conductivities are listed with references.

Table 6.5-1: thermal conductivity

Material	Thermal Conductivity (W/(mK))	Reference
Air (@ 85°C)	0.0305	[16]
Silicon rubber sealant	0.19	[17], [18]*
LUBRIGLAS® AP-108-03	0.03	Measure
Sandvik 1802® (magnetic inox steel)	22.5	[7]
ARMCO® Pure Iron	73.15	[6]

*The products selected in [17] and [18] are typical silicon rubber sealant.

The measurement of LUBRIGLAS® AP-108-03 thermal conductivity has been made in the following way. A test has been performed with a reel of a solenoid with coils winded. Above windings, a sheet of LUBRIGLAS® AP 108-03 has been applied. The scheme of the test is represented in Figure 6.5-1.

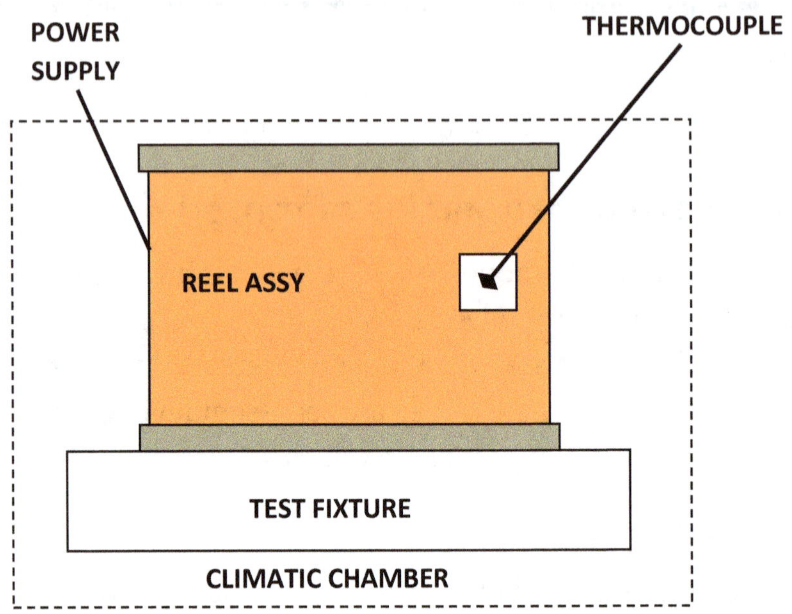

Figure 6.5-1: LUBRIGLAS® thermal conductivity measurement scheme.

We apply a DC tension and wait for the equilibrium condition. At equilibrium we record the current I, the resistance R and the external temperature $T_{LUBRIGLAS}$, that has been measured with a thermocouple applied on the external surface of the fabric. Current has been measured with an amperometer, while resistance has been computed with the Ohm law.

Now we can write power dissipated by the reel assy at equilibrium, assuming that heat is totally exchanged in the radial direction. This power dissipated by Joule effect is equal to thermal power dissipated through LUBRIGLAS® fabric:

$$P = RI^2 = \frac{\lambda 2\pi Y}{\ln\left(\frac{r_e}{r_i}\right)}\left(T_{COIL} - T_{LUBRIGLAS}\right) \tag{6.5.1}$$

In the (6.5.1), we can find the following parameters:

- λ = thermal conductivity of LUBRIGLAS®
- Y = windings height
- r_e = external radius of LUBRIGLAS®
- r_i = internal radius of LUBRIGLAS®
- T_{COIL} = average coil temperature

Average coil temperature is indirectly measured from the knowledge of R_0 (measure with an Ohmmeter) and reference temperature T_0:

$$T_{COIL} = \left(\frac{R}{R_0} - 1 \right) \frac{1}{\alpha} + T_0 \tag{6.5.2}$$

Expressing thermal conductivity from (6.5.1):

$$\lambda = \frac{RI^2 \ln\left(\dfrac{r_e}{r_i}\right)}{2\pi\left(T_{COIL} - T_{LUBRIGLAS}\right)} \tag{6.5.3}$$

If we call "t" the thickness of LUBRIGLAS®, we can write:

$$\lambda = \frac{RI^2 \ln\left(\dfrac{r_e}{r_e - t}\right)}{2\pi\left(T_{AVV} - T_{LUBRIGLAS}\right)} \tag{6.5.4}$$

We substitute the following data in (6.5.4):

- T_0 = 18°C
- V = 28 VDC
- R_0 = 30.94 Ω
- Y = 32.7 mm
- t = 0.12 mm
- r_e = 16.35 mm
- I = 0.549 mm
- $T_{LUBRIGLAS}$ = 169 °C

We obtain that:

$$R = V/I = 51.0 \ \Omega$$

$$T_{COIL} = 184.26 \ °C$$

$$\lambda = 0.03175 \ W/(m \ K)$$

We underline that this value is not of the same order of magnitude of the one measured in [19] for a similar PTFE pre-impregnated fiberglass fabric. In [19] a value of $\lambda \approx 0.2$ W/(m K) has been measured (the proper value depends from the kind of fabric). The reason of this difference stays in the fact that air is trapped in the fiberglass and between the LUBRIGLAS® fabric and the surface where it is applied. So the value we have measured is considered as accurate and reliable for solenoid applications.

<u>Density</u>

This property is important to compute total mass M, used in the transitory model, from dimensional data taken from solenoid sizing procedure.

Table 6.5-2: Density of materials.

Material	Density (g/cm³)	Reference
Air (@ 25°C)	0.001225	[20]
LUBRIGLAS® AP-108-03	2.02	Measure
Sandvik 1802® (magnetic inox steel)	7.85	[7]
ARMCO® Pure Iron	7.85	[6]
AISI 316	7.85	[21]
Aluminium	2.90	[22]
Copper	8.93	[22]
Magnetic wire enamel**	1.37	-

**The material is a polyesterimide over coated with polyamide-imide. The value in Table 6.5-2 is typical of a generic polyester kind material.

The measurement of LUBRIGLAS® AP-108-03 density has been made with a piece of fabric with the following dimensions:

- l_1 = 75 mm (side 1)
- l_2 = 44 mm (side 2)
- s = 0.12 mm (thickness)

The volume is V = 75*44*0.12=396 mm^3 = 0.396 cm^3.

The measured weight is 0.8 g. The density is:

$$\rho = \frac{m}{V} = \frac{0.8}{0.396} g / cm^3 = 2.02 g / cm^3$$

<u>Specific heat</u>

This property is important to compute total equivalent specific heat C, used in the transitory model.

Table 6.5-3: Specific heat of materials.

Material	Specific heat (J/(kg K))	Reference
Air (@ 25°C)	1006	[20]
LUBRIGLAS® AP-108-03	1120	Indirect method
Sandvik 1802® (magnetic inox steel)	500	[7]
ARMCO® Pure Iron	450	[6]
AISI 316	500	[21]
Aluminium	900	[22]
Copper	385	[22]
Magnetic wire enamel***	1200	[23]

***The material is a polyesterimide overcoated with polyamide-imide. The value in Table 6.5-3 is typical of a generic polyester kind material.

The computation of specific heat of LUBRIGLAS® AP-108-03 has been made considering the fact that it is a composite material typically composed by: 60% of PTFE and 40% of fiberglass. So we can write:

$$c_{P\,LUBR} = 0.6 c_{P\,PTFE} + 0.4 c_{P\,FIBERGLASS} \tag{6.5.5}$$

Typical values for fiberglass and PTFE are:

$c_{P\,PTFE}$ = 1300 J/(kg K)

$c_{P\,FIBERGLASS}$ = 850 J/(kg K)

Therefore, we have:

$$c_{P\,LUBR} = 1090 \frac{J}{kg \cdot K}$$

6.6. Computation of total equivalent specific heat

The specific heat C in (6.3.1) is computed in the following way:

$$C = \frac{\sum_i V_i C_i}{\sum_i V_i} \tag{6.6.1}$$

where V_i and C_i are respectively the volumes and specific heat of internal parts of solenoid (keeper, core, reel, insulant fabric and windings). The volumes of internal parts are typically computed in the following way:

$$V_i = \frac{\pi \left(D_e^2 - D_i^2\right)}{4} Y \tag{6.6.2}$$

where D_e and D_i are respectively the external and internal diameter of the i-th layer.

The computation of the windings zone is made in a different way. We have to compute the volume of the enamel, of copper and of air trapped in the windings. First of all, we compute the average volume of the wire as:

$$V_W = \frac{\pi D_m^2}{4} L \qquad (6.6.3)$$

where D_m is the average diameter of wire computed as:

$$D_m = \frac{D_W + D_{WINT}}{2} \qquad (6.6.4)$$

The volume of copper is:

$$V_C = \frac{\pi D_{IW}^2}{4} L \qquad (6.6.5)$$

Therefore, we can write the volume of enamel as:

$$V_E = V_W - V_C \qquad (6.6.6)$$

In order to compute the volume of trapped air we first have to compute the total volume of winding zone:

$$V_{W/A} = \frac{\pi}{4}\left[(D_1 + 2X)^2 - D_1^2\right]Y \qquad (6.6.7)$$

Therefore, we can write the volume of air as:

$$V_A = V_{W/A} - V_W \qquad (6.6.8)$$

6.7. Heat exchange coefficient K

All the computation of the thermal performances of a solenoid depends strictly from the heat exchange coefficient with external ambient K introduced in (6.2.6). This coefficient influences the value of H in (6.2.7) and it is the most important parameter of the model. However, this parameter is rarely known. It depends from solenoid installation and orientation and these data are often unknown. Moreover, if it is possible to know the solenoid installation is practically impossible to determine in a theoretical way its value, because of the complexity of the heat exchange problem with external ambient, involving radiation and convection.

The solution is to define a standard position and installation of the solenoid and perform measurements of the coefficient. In this way, it is possible to make an estimation of thermal performances and to compare the behaviour of different solenoids. It is easy to understand that this approach needs to record results of thermal performances in order to be more confident with the model and computations.

The standard position and installation is shown in the scheme of Figure 6.7-1.

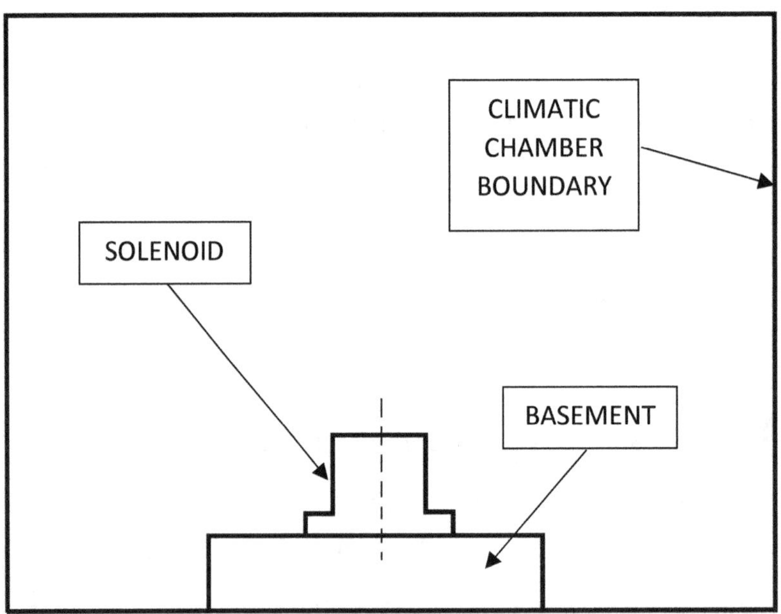

Figure 6.7-1: Thermal performances measurement scheme.

The solenoid is installed on a basement in a climatic chamber with constant temperature control. The solenoid axis is vertical.

Before the test, electrical resistance R_0 at reference temperature T_0 and ambient temperature T_{AMB} are recorded. The solenoid is energized with a known voltage; current absorption is recorded with an amperometer; outer diameter temperature T_4 is recorded with a thermocouple. From current absorption it is possible to monitor coil resistance and so temperature from the following formula:

$$T = T_0 + \frac{1}{\alpha}\left(\frac{R}{R_0} - 1\right) \tag{6.7.1}$$

If coil temperature does not exceed maximum temperature (typically 200 °C), it is possible to perform measurement until equilibrium. We can write dissipated power at equilibrium as:

$$W = VI_E = K2\pi r_4 Y\left(T_{4_E} - T_{AMB}\right) \tag{6.7.2}$$

where T_{4E} and I_E are respectively outer diameter temperature and absorbed current at equilibrium. Therefore, we can explicit K:

$$K = \frac{VI_E}{2\pi r_4 Y\left(T_{4_E} - T_{AMB}\right)} \tag{6.7.3}$$

In the following figures, we can observe that if we use a correct heat exchange coefficient K the transitory model is very accurate.

We want to underline that transitory model for the computation of external temperature of the solenoid has not been made. In fact, usually we interested to coils temperature, in order to avoid damages of the magnet wire.

However, in some applications, it is useful to know external temperature. For example, if the solenoid is installed in a position where flammable gas is present, it is important that solenoid external temperature is not equal or higher than ignition temperature of gas. In that case, it is important to evaluate external temperature.

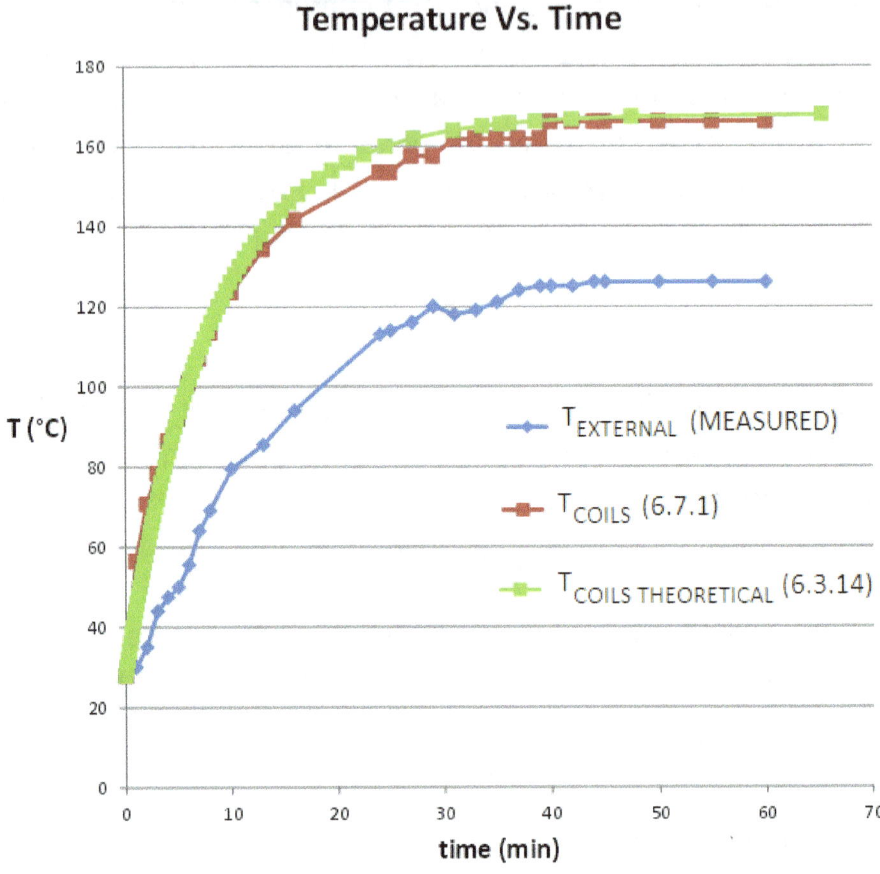

Figure 6.7-2: Theoretical and experimental temperatures comparison.

If we evaluate the resistance of the coil at the temperature calculated with (6.3.14), we are able to evaluate the current absorption with time.

Even in this case, it is possible to appreciate the accuracy of the thermal model presented here. Of course, the heat exchange coefficient K will be very different for different geometry and installation. Therefore, the computational technique described here shall be considered as a reference for evaluating thermal performances of the device.

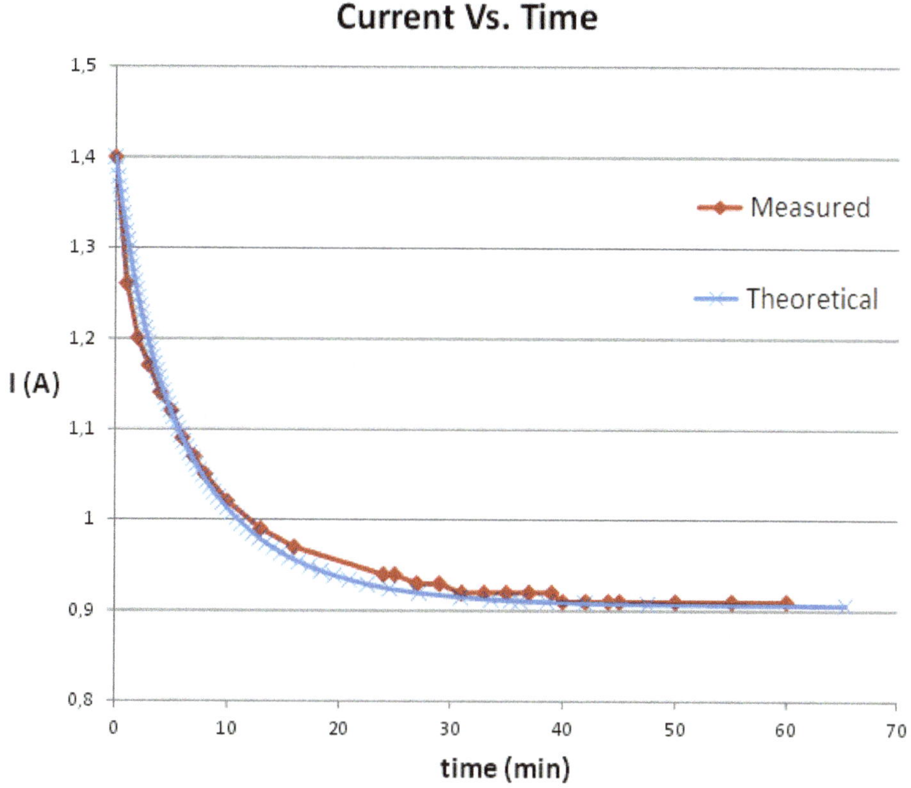

Figure 6.7-3: Theoretical and experimental absorbed currents comparison.

Chapter 7

Advanced Calculations

7.1. Electro-mechanical model

In this section, we will describe advanced computations in order to understand in a deeper way solenoid behavior. At this point, we have described how to compute magnetic force and inductance but we did not focus the attention to actuation times and dynamic behavior.

Now we describe the electro-mechanical model of the solenoid. We have to write two equations: the first one describes the evolution of coils current with time; the second one describes the evolution of the position of the solenoid keeper with time.

First equation: electrical description

The electrical scheme of the solenoid is represented in Figure 7.1-1, where we can see the voltage source, coils resistance, inductance and switch.

We can write the equation of evolution of the current:

$$L(x, I)\frac{dI(t)}{dt} = V(t) - RI(t) \qquad (7.1.1)$$

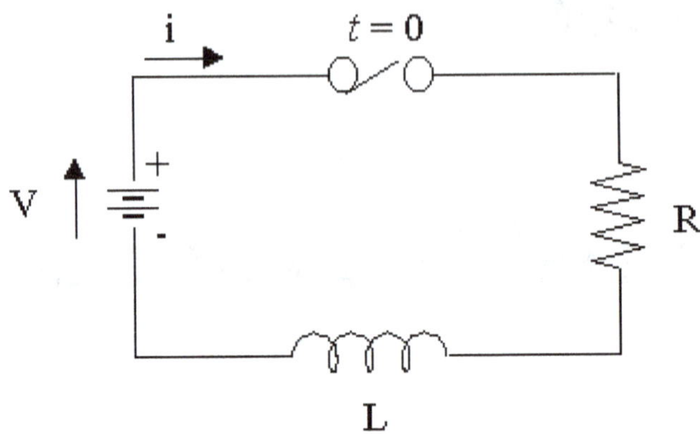

Figure 7.1-1: Solenoid electrical scheme.

In this equation, we can see that inductance depends from current and position of the keeper "x" (air gap of the solenoid), as we have seen in chapter 4. In order to calculate the evolution of current with time, we analyze the case of an inductor with constant value inductance and a voltage supply represented by the step function (see Figure 7.1-2).

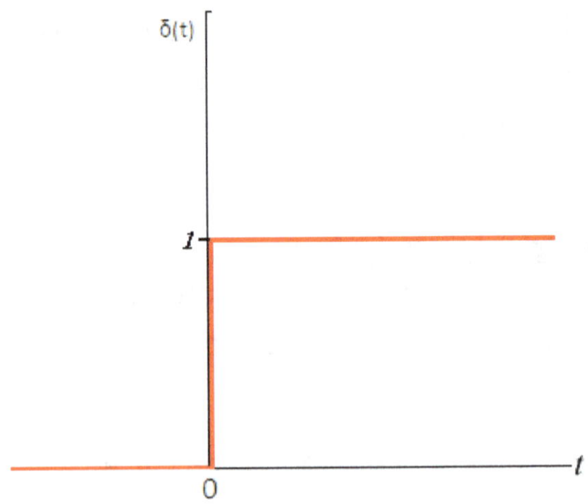

Figure 7.1-2: Step function.

If we denote with δ the unit step function and with V_s the stationary voltage, we can write:

$$L\frac{dI(t)}{dt} = V_s\delta(t) - RI(t) \qquad (7.1.2)$$

The solution of this equation is:

$$\begin{cases} I(t) = I_S(1 - e^{-\frac{t}{\tau}}) \\ I_s = \frac{V_S}{R} \\ \tau = \frac{L}{R} \end{cases} \qquad (7.1.3)$$

where I_S is the stationary current and τ is the time constant of the system. The evolution of current with time is represented in Figure 7.1-3.

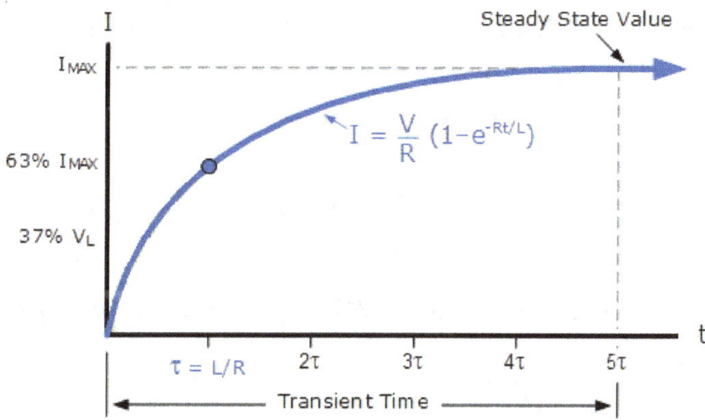

Figure 7.1-3: Current Vs. Time.

It is easy to understand that the ratio between inductance and resistance determines the dynamic behavior of the system.

For such a system, this analysis can be sufficient. In order to better understand the more general case (non-constant inductance), now we will see in detail a very powerful instrument which is very useful for understanding dynamic behavior of a solenoid: the study of the frequency response.

Frequency Response

A generic linear system of the first order can be represented as:

$$\dot{x} = Ax + Bu$$

$$x = \begin{cases} x_1 \\ x_2 \\ \dots \\ x_N \end{cases}$$

$$u = \begin{cases} u_1 \\ u_2 \\ \dots \\ u_M \end{cases}$$

(7.1.4)

Where the array "x" represents the independent variables and the array "u" represents the inputs. In solving differential equations, it is useful to define the Laplace Transform:

$$F(s) = \int_0^{+\infty} f(t)e^{-st} dt \qquad (7.1.5)$$

If we apply this operation to the system in (7.1.4), we can write the following equation:

$$x(s) = \Phi(s)x_0 + W(s)u(s)$$
$$\Phi(s) = (sI - A)^{-1} \qquad (7.1.6)$$
$$W(s) = (sI - A)^{-1} B$$

Where x_0 is the initial state of the system. The $W(s)$ is called Transfer Function. This function is very important because it can be demonstrated that, if the input is sinusoidal, after a transitory period, the permanent response is sinusoidal with an amplitude equal to this function:

$$u(t) = u_0 \sin(\omega t)$$
$$x(t) = u_0 M(\omega)\sin(\omega t + \phi(\omega))$$
$$M(\omega) = |W(j\omega)| \tag{7.1.7}$$
$$\phi(\omega) = \angle W(j\omega)$$

Therefore, it can be useful to analyze this function, in particular the modulus M. In the case of equation (7.1.2), written as (7.1.4), the transfer function is:

$$W(s) = \frac{1}{R}\frac{1}{1+\tau s}$$
$$M(\omega) = \frac{1}{R^2}\frac{1}{\sqrt{1+\omega^2\tau^2}} \tag{7.1.8}$$

A good and simple representation of the transfer function is done by the so called Bode Diagram. It consists in the representation of the following:

$$|W(j\omega)|_{dB} = 20\log_{10}|W(j\omega)| \tag{7.1.9}$$

In our case:

$$M(\omega)_{dB} = 20\log_{10}\left(\frac{1}{R^2}\right) - 20\log\sqrt{1+\omega^2\tau^2} \tag{7.1.10}$$

The first term of (7.1.10) is just a constant:

$$C_{dB} = 20\log_{10}\left(\frac{1}{R^2}\right) \tag{7.1.11}$$

The second term need to be described. If we consider very low pulsations ω this term is very close to be zero. For higher values of ω the (7.1.10) can be written as:

$$\varphi_a = C_{dB} - 20\log\sqrt{1+\omega^2\tau^2} \approx C_{dB} - 20\log\sqrt{\omega^2\tau^2} =$$
$$= C_{dB} - 20\log\omega\tau = C_{dB} - 20\log\tau - 20\log\omega \tag{7.1.12}$$

Equation (7.1.12) is the asymptotic response and it represents a straight line with inclination of 20dB/dec that crosses the C_{dB} value for $\omega=1/\tau$. This value is called Rupture Pulsation because, for pulsations higher than it, the system has a negligible response. So a solenoid can be seen as a low-pass filter: the rupture value is the cut-off frequency:

$$f_{CU} = \frac{R}{2\pi L} \tag{7.1.13}$$

For a solenoid with R=50 Ω and L=0.5 H (typical values of a solenoid valve) f_{CU} = 15.9 Hz and the rupture pulsation is ω_R = 100 rad/s.

If we observe the Bode diagram of such a solenoid in Figure 7.1-4, we can see that for high frequency the solenoid has a response practically equal to zero. Lines in red are the asymptotic responses of the system. Their intersection point is the rupture pulsation.

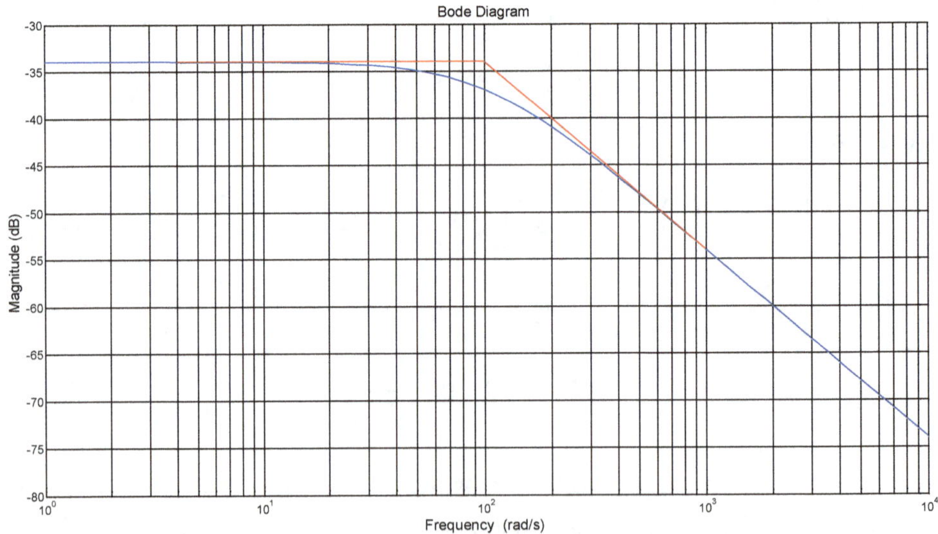

Figure 7.1-4: Bode diagram of a typical solenoid.

In the semi-logarithmic diagram, for a decade before and after the rupture pulsation, the difference between the asymptotic values and the effective diagram is negligible:

$$\left[M_{dB} - \varphi_a\right]_{\omega=\frac{1}{10\tau}} = C_{dB} - 20\log_{10}\sqrt{1+\frac{1}{100}} - (C_{dB}) \approx -0.043 dB$$

$$\left[M_{dB} - \varphi_a\right]_{\omega=\frac{10}{\tau}} = C_{dB} - 20\log_{10}\sqrt{1+100} - (C_{dB} - 20 dB) \approx -0.043 dB$$

$$(7.1.14)$$

The maximum difference stays for the rupture pulsation:

$$\left[M_{dB} - \varphi_a\right]_{\omega=\frac{1}{\tau}} = C_{dB} - 20\log_{10}\sqrt{2} - (C_{dB}) \approx -3 dB \qquad (7.1.15)$$

This analysis can be done for a more complex system with a higher number of variables and input. The only requirement is that the system must be linear. Now we will describe the complete electro-mechanical system of a solenoid, which is nonlinear. After that, we will describe the process of linearization and analysis of this complete system.

Second equation: mechanical description

The second equation is the mechanical description of the solenoid. As we have seen before, the solenoid consists in a moving keeper that is moved by magnetic force. This force acts against keeper inertia, back spring return force and an external load. This load is typically a pressure force for a solenoid valve. In the case of an actuator, the external load depends from application. Therefore, it is easy to understand that a solenoid can be modeled as classical mass, spring and damper system (Fig. 5).

In the scheme of Figure 7.1-5, there are the following quantity which characterize the mechanical model:

- m: is the mass of the moving parts of the solenoid
- k: is the stiffness of the return spring
- x_0: is the preload elongation of the spring
- b: is the damping coefficient of the system

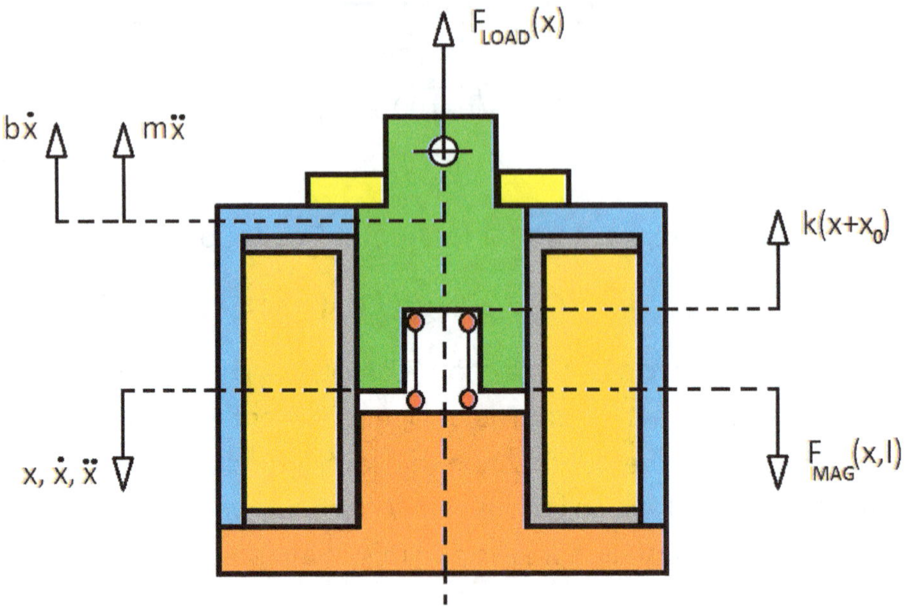

Figure 7.1-5: Solenoid mechanical scheme.

This last quantity is usually unknown in every dynamic system in engineering problem. There are different methods for the representation of this quantity. A typical representation considers the damping coefficient as a percentage of critical damping. Critical damping is the value that does not gives the possibility to the system to oscillate. Its value is:

$$b_{CR} = 2\sqrt{km} \qquad (7.1.16)$$

A typical choice is:

$$b = 0.05 b_{CR} \qquad (7.1.17)$$

Moreover, we want to underline that the quantity F_{LOAD} and F_{MAG} are known. They have to be computed before resolving the differential equation of the system that is:

$$m\ddot{x} + b\dot{x} + k(x + x_0) = F_{MAG}(x, I) - F_{LOAD}(x) \qquad (7.1.18)$$

Solenoid system and solving

Now we can write all the system of equations:

$$\begin{cases} L(x,I)\dfrac{dI(t)}{dt} = V(t) - RI(t) \\ m\ddot{x} + b\dot{x} + k(x + x_0) = F_{MAG}(x,I) - F_{LOAD}(x) \end{cases} \tag{7.1.19}$$

As we can see from (7.1.19), this system is not linear because inductance and magnetic force are not linear and, above all, they are not constant. The only way of solving this system is to use a computer approach. In particular a software as Simulink© can be used.

A short description of the solving method is needed. The following steps has to be done:

1. <u>Preprocess</u>: inductance and magnetic force functions have to be computed. For magnetic force we will obtain a function similar to Figure 7.1-6.

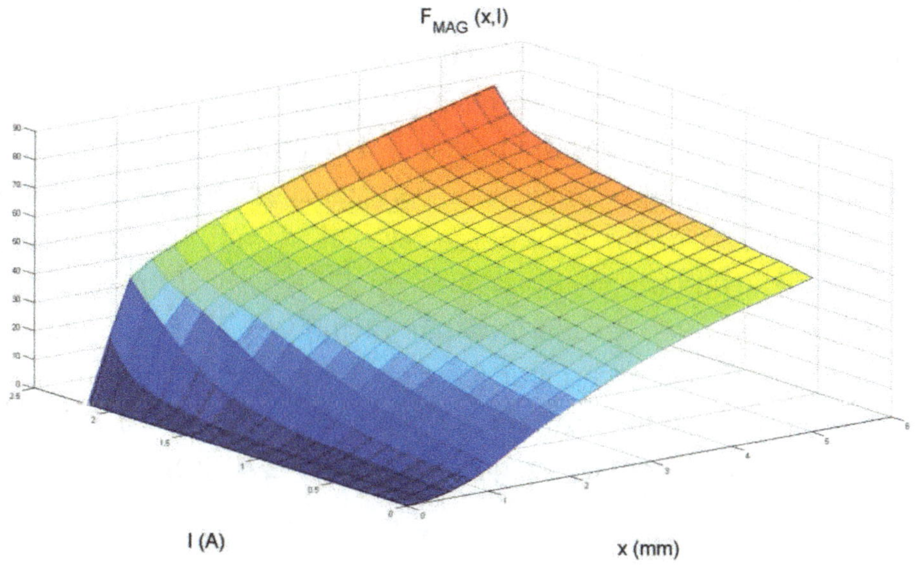

Figure 7.1-6: Magnetic force representation.

2. <u>Problem definition</u>: mechanical and electrical characteristics of the solenoid will be defined. We will choose the following quantity:
    - Mass of moving parts;
    - Spring characteristics: stiffness and preload;
    - Initial position of moving parts;
    - Resistance of the coil;
    - Tension as function of time;
    - External load ad function of x (in some case it can depend even from time).
3. <u>Simulation of the system</u>: a Simulink© model has to be defined. In Figure 7.1-7 a typical model is represented.

After the simulation is completed, we can analyze the solution and plot the variables "x" and "I" as function of time (see Figure 7.1-8 and Figure 7.1-9). The simulation has been done with a voltage represented by a step function.

As we can see from Figure 7.1-8, the current behavior is similar to Figure 7.1-3, but in this case we have computed the complete model of the solenoid. It is interesting to see in detail the position graph. In particular, we can see two important characteristics: the first one refers to the fact that the graph is cut at the value of 5.0 mm. This is because this value is the end of the stroke of the moving parts of the solenoid. The second one refers to the fact that, at the very first moments of voltage application, the magnetic force does not overcome spring load and external load. So the position of the moving parts is zero. This fact can be seen with a zoom of the graph (see Figure 7.1-10).

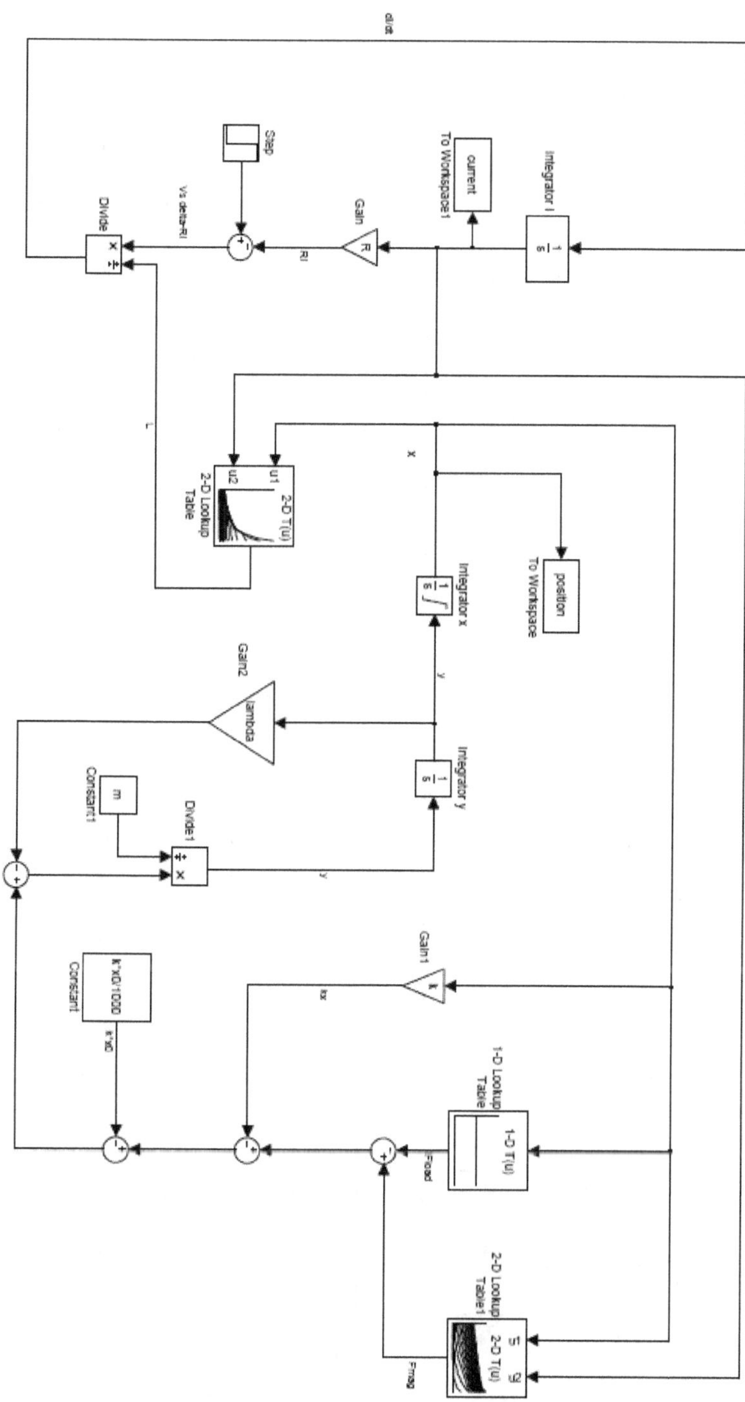

Figure 7.1-7: Model representation.

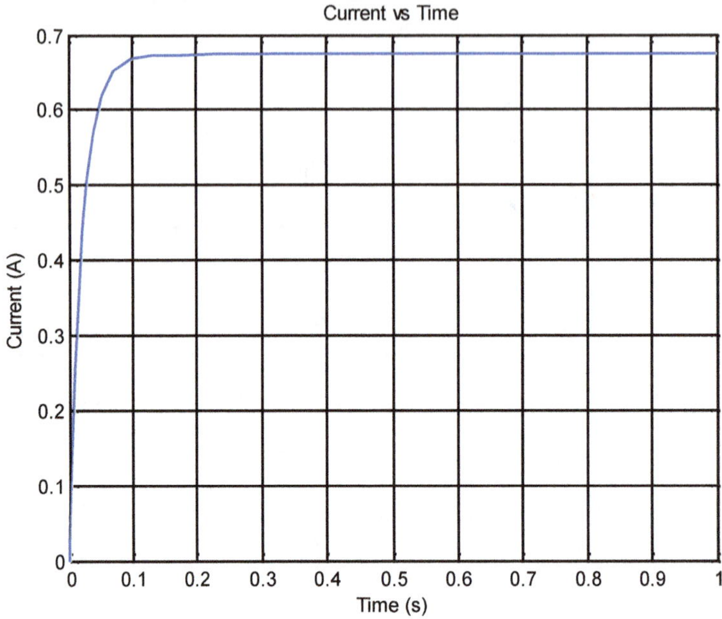

Figure 7.1-8: Current vs. Time.

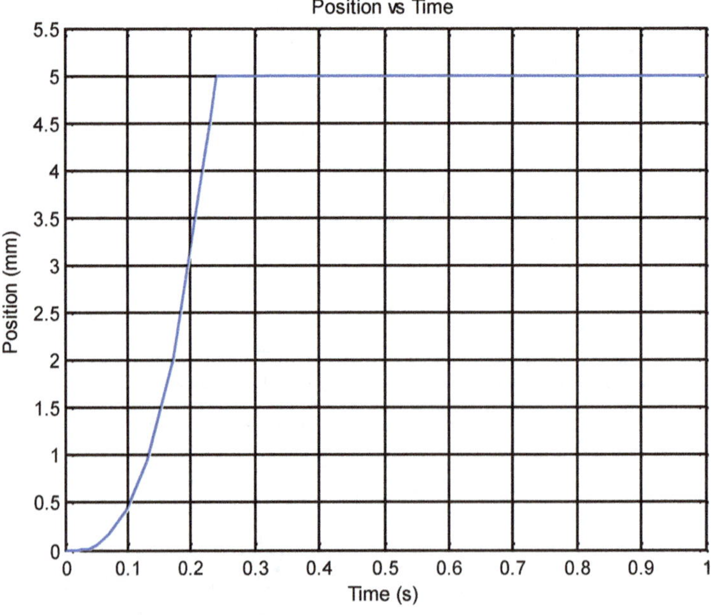

Figure 7.1-9: Position vs. Time.

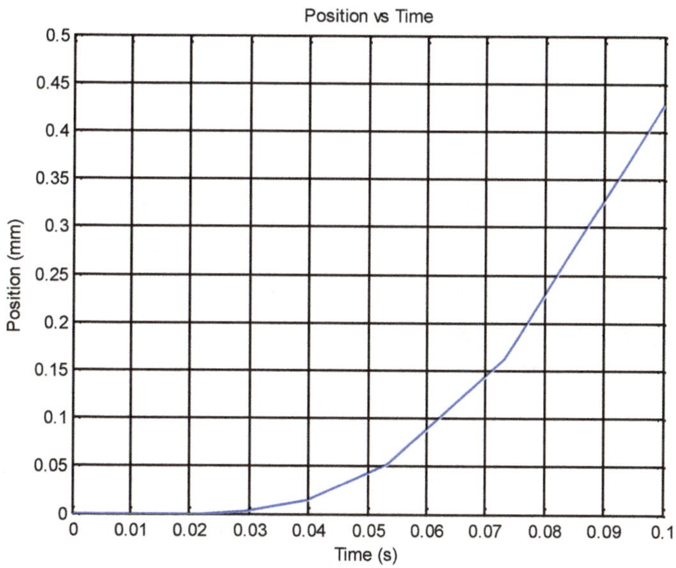

Figure 7.1-10: Position vs. Time zoomed.

Frequency response of the electro-mechanical model

The system (7.1.19) is nonlinear. Therefore, the frequency response of this system does not exist. However, we can compute the frequency response of the system nearby an equilibrium point. In order to do that, we have to linearize the system.

First of all we write the system with the representation of (7.1.4):

$$\begin{cases} \dot{x} = y \\ \dot{y} = \dfrac{F_{MAG}(x,I) - F_{LOAD}(x) - k(x+x_0) - by}{m} \\ \dot{I} = \dfrac{V(t) - RI}{L(x,I)} \end{cases} \qquad (7.1.20)$$

In the moment that magnetic force equals external load and spring load we can write:

$$\begin{cases} \ddot{x} = 0 \\ \dot{x} = 0 \\ kx_0 = F_{MAG}(0, I_E) - F_{LOAD}(0) \end{cases}$$ (7.1.21)

This is mechanical equilibrium point. If we denote as I_E the current value that allows this equilibrium, which is known from (7.1.21), we can compute exactly the voltage value able to guarantee not only this equilibrium but also the electrical one:

$$V_E = RI_E$$ (7.1.22)

Therefore, we can write the equilibrium point of the system as:

$$x = x_E = 0$$
$$y = y_E = 0$$
$$I = I_E$$ (7.1.23)
$$V = V_E = RI_E$$

Now we will use the formalism of (7.1.4) and we will define new variables:

$$x_1 = x - x_E = x$$
$$x_2 = y - y_E = y$$
$$x_3 = I - I_E$$ (7.1.24)
$$u = V - V_E$$

Those variables represent the deviations from equilibrium point. The system is generically written as:

$$\dot{x}_1 = f_1(X, u)$$
$$\dot{x}_2 = f_2(X, u)$$ (7.1.25)
$$\dot{x}_3 = f_3(X, u)$$

The linearized representation of the system is based on Taylor decomposition of the first order:

$$\left\{\begin{array}{c} \dot{x}_1 \\ \dot{x}_2 \\ \dot{x}_3 \end{array}\right\} = \left[\begin{array}{ccc} \dfrac{\partial f_1}{\partial x_1} & \dfrac{\partial f_1}{\partial x_2} & \dfrac{\partial f_1}{\partial x_3} \\[2ex] \dfrac{\partial f_2}{\partial x_1} & \dfrac{\partial f_2}{\partial x_2} & \dfrac{\partial f_2}{\partial x_3} \\[2ex] \dfrac{\partial f_3}{\partial x_1} & \dfrac{\partial f_3}{\partial x_2} & \dfrac{\partial f_3}{\partial x_3} \end{array}\right]_E \left\{\begin{array}{c} x_1 \\ x_2 \\ x_3 \end{array}\right\} + \left\{\begin{array}{c} \dfrac{\partial f_1}{\partial u} \\[2ex] \dfrac{\partial f_2}{\partial u} \\[2ex] \dfrac{\partial f_3}{\partial u} \end{array}\right\}_E u \qquad (7.1.26)$$

The subscript "E" refers to the fact that derivatives must be computed in the equilibrium position. If we look at (7.1.26), we can recognize the representation (7.1.4) of the system. Now we can calculate the derivative in (7.1.26):

$$f_1 = y$$

$$\left.\frac{\partial f_1}{\partial x_1}\right|_E = 0 \ ; \quad \left.\frac{\partial f_1}{\partial x_2}\right|_E = 1 \ ; \quad \left.\frac{\partial f_1}{\partial x_3}\right|_E = 0 \ ; \quad \left.\frac{\partial f_1}{\partial u}\right|_E = 0$$

$$f_2 = \frac{1}{m}\left[F_{MAG} - F_{LOAD} - k(x + x_0) - by\right]$$

$$\left.\frac{\partial f_2}{\partial x_1}\right|_E = \frac{1}{m}\left(\left.\frac{\partial F_{MAG}}{\partial x}\right|_{\substack{x=0 \\ I=I_E}} - \left.\frac{\partial F_{LOAD}}{\partial x}\right|_{x=0} - k\right) \ ; \quad \left.\frac{\partial f_2}{\partial x_2}\right|_E = -\frac{b}{m} \ ;$$

$$\left.\frac{\partial f_2}{\partial x_3}\right|_E = \frac{1}{m}\left.\frac{\partial F_{MAG}}{\partial I}\right|_{\substack{x=0 \\ I=I_E}} \ ; \quad \left.\frac{\partial f_2}{\partial u}\right|_E = 0$$

$$f_3 = \frac{u - RI}{L}$$

$$\left.\frac{\partial f_3}{\partial x_1}\right|_E = \left[(u - RI)\left(-\frac{1}{L^2}\right)\frac{\partial L}{\partial x}\right]_E = 0 \ ; \quad \frac{\partial f_3}{\partial x_2} = 0 \ ;$$

$$\left.\frac{\partial f_3}{\partial x_3}\right|_E = \left[-\frac{R}{L} + (u - RI)\left(-\frac{1}{L^2}\right)\left(\frac{\partial L}{\partial I}\right)\right]_E = -\left.\frac{R}{L}\right|_{\substack{x=0 \\ I=I_E}} \ ; \quad \left.\frac{\partial f_3}{\partial u}\right|_E = \left.\frac{1}{L}\right|_{\substack{x=0 \\ I=I_E}}$$

From equations above, we can build the linear system in the representation of (7.1.4) and so we can define the matrix "A" and "B":

$$A = \begin{bmatrix} 0 & 1 & 0 \\ \dfrac{1}{m}\left(\dfrac{\partial F_{MAG}}{\partial x}\bigg|_{\substack{x=0 \\ I=I_E}} - \dfrac{\partial F_{LOAD}}{\partial x}\bigg|_{x=0} - k \right) & -\dfrac{b}{m} & \dfrac{1}{m}\dfrac{\partial F_{MAG}}{\partial I}\bigg|_{\substack{x=0 \\ I=I_E}} \\ 0 & 0 & -\dfrac{R}{L}\bigg|_{\substack{x=0 \\ I=I_E}} \end{bmatrix}$$

$$(7.1.27)$$

$$B = \begin{bmatrix} 0 \\ 0 \\ -\dfrac{1}{L}\bigg|_{\substack{x=0 \\ I=I_E}} \end{bmatrix} \qquad (7.1.28)$$

Now we are able to compute the Transfer Function $W(s)$ in (7.1.6) and perform the analysis.

7.2. Electromagnetic compatibility

A solenoid is an electromagnetic device. So it is affected by electromagnetic disturbances from external actions. These actions are typically due to electrical devices in the nearby of the solenoid. Electromagnetic compatibility (EMC) is the branch of electrical engineering that studies and analyses these kind of actions, in order to evaluate if the device that has been analyzed is robust to external disturbances. RTCA-DO 160G Section 19 (see [24]) details this kind of analyses and tests in the aerospace industry.

We will perform an analysis in order to demonstrate the compatibility of a solenoid valve to the test of Table 7.2-1:

Table 7.2-1: EMC test list.

Test	Category ZC levels
Magnetic Fields induced into equipment	20A rms at 400Hz
Electrical fields induced into equipment	170V rms at 400Hz
Magnetic fields induced into interconnecting cables	IxL = 30Am at 400Hz reducing to 0.8 15kHz
Electrical Fields induced into interconnecting cables	VxL = 1800Vm from 380 to 420Hz
Spikes induced into interconnecting cables	L=3.0m

Magnetic Fields induced into equipment

The test consists in a circuit positioned nearby the unit under test, in our case the solenoid. The AC current of the circuit will produce an induced voltage variable with time inside the solenoid coil. This voltage shall be lower of the minimum voltage required for energizing the solenoid (see Figure 7.2-1).

The analysis that we will perform is very conservative: we will consider the solenoid without its external magnetic case. So there won't be the external ferromagnetic material that would concentrate external magnetic flux thanks to its high magnetic permeability; so all the flux lines will be concatenated with the coil.

To simplify the calculation, we will consider the coil section not as a circle but as a square with a side of 2R (where R is the external radius of the coils). This is a conservative hypothesis that guarantees a higher external flux across the solenoid. Now we compute this flux following Figure 7.2-2.

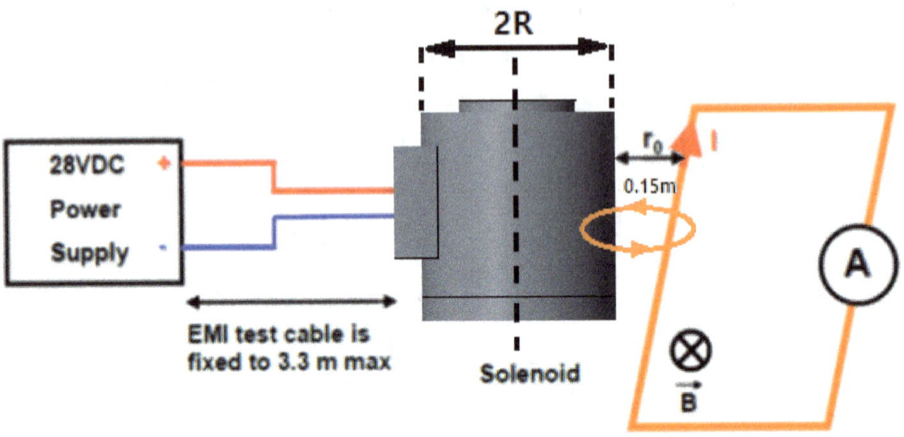

Figure 7.2-1: Magnetic Fields induced into equipment: test set up.

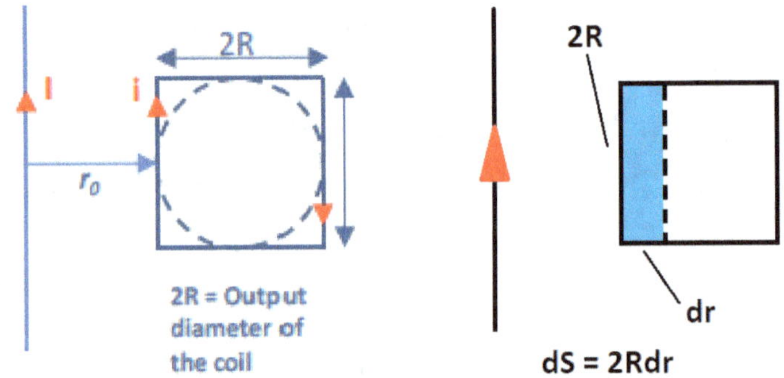

Figure 7.2-2: Magnetic Fields induced into equipment: calculation scheme.

We remind the Biot-Savard law for the magnetic induction field of a wire:

$$B = \frac{\mu_0 I}{2\pi r_0} \qquad (7.2.1)$$

We want to underline that we do not introduce relative permeability because air and copper are amagnetic. The flux:

$$\phi = \int BdS = \int \frac{\mu_0 I}{2\pi r} dS = \frac{\mu_0 I \cdot 2R}{2\pi} \int_{r_0}^{r_0+2R} \frac{dr}{r} = \frac{\mu_0 I \cdot R}{\pi} \ln\left(\frac{r_0 + 2R}{r_0}\right)$$

$$(7.2.2)$$

According to the Faraday-Neumann-Lenz law, the induced voltage is:

$$e = -N_{TURNS} \frac{\partial \phi}{\partial t} = -N_{TURNS} \frac{\mu_0 R}{\pi} \ln\left(\frac{r_0 + 2R}{r_0}\right) \frac{\partial I}{\partial t} \qquad (7.2.3)$$

We substitute the current:

$$I(t) = \sqrt{2} I_{RMS} \sin(2\pi f t) \qquad (7.2.4)$$

We obtain:

$$e = -N_{TURNS} \mu_0 I_{RMS} 2\sqrt{2} R f \ln\left(\frac{r_0 + 2R}{r_0}\right) \cos(2\pi f t) \qquad (7.2.5)$$

So the maximum value of the induced voltage is:

$$\left|e_{MAX}\right| = N_{TURNS} \mu_0 I_{RMS} 2\sqrt{2} R f \ln\left(\frac{r_0 + 2R}{r_0}\right) \qquad (7.2.6)$$

For a typical solenoid valve application:

I_{RMS} = 20 A;

r_0 = 0.15 m;

R = 14 mm;

N = 2500;

f = 400 Hz.

Using (7.2.6) the maximum induced voltage is e_{MAX} = 170 mV. For such a solenoid, the minimum energizing voltage is about 4 – 5 V and so this test does not affect solenoid functioning.

Electric Fields induced into equipment

This test consists in energizing the solenoid with an alternate voltage with a typical frequency of 400 Hz. We does not need to perform more analysis because we have already demonstrate that a solenoid is a low-pass filter and it is not affected by this test.

Magnetic Fields induced into interconnecting cables

This test consists in the evaluation of the effects induced on the interconnecting cables of the unit under test by another circuit (see Fig. 12).

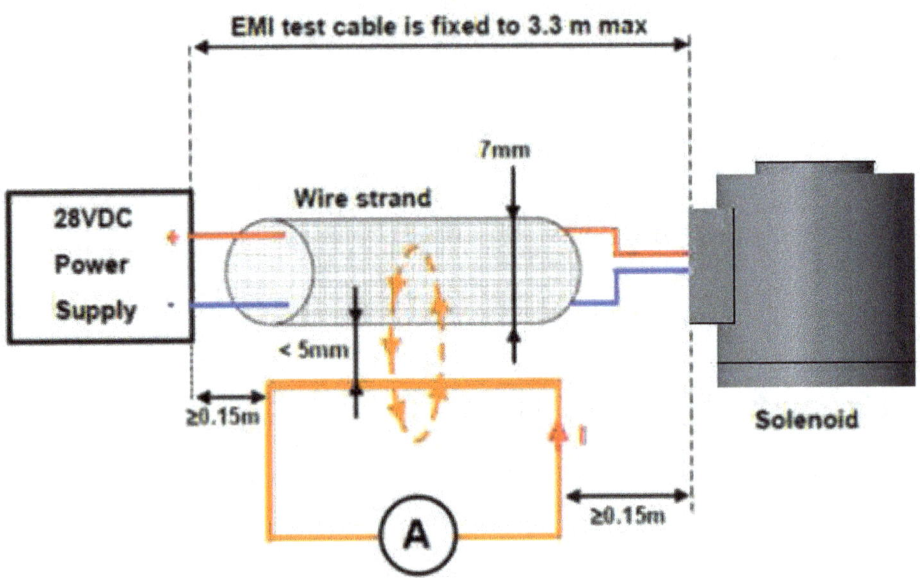

Figure 7.2-3: Magnetic Fields induced into interconnecting cables: test set up.

We take in account the following conservative assumptions:

- The supply conductors are not twisted.
- No shielding for the supply wires.
- There is a distance between the supply wires disturbed by the magnetic field (they form a coil).

We have two circuits: the "C_1" circuit is the disturbing one; the "C_2" is the supply circuit (see Figure 7.2-4).

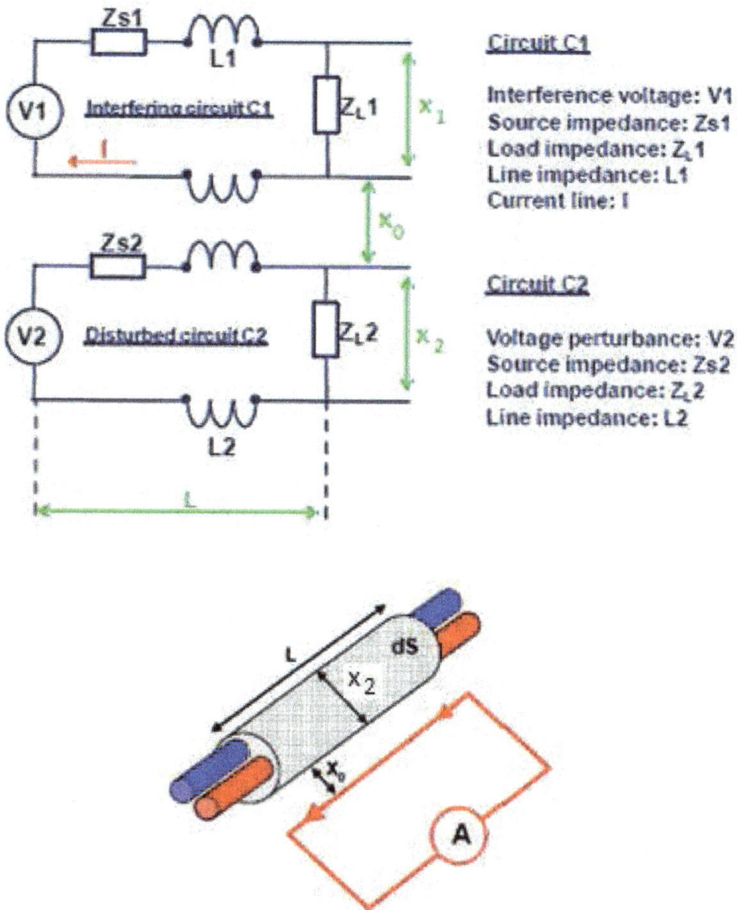

Figure 7.2-4: Magnetic Fields induced into interconnecting cables: calculation scheme.

First, we have to compute the induced voltage V_2 as we have done before. The disturbed circuit is a coil with an area $S=Lx_2$. So we can write:

$$dS = Ldr \qquad (7.2.7)$$

The magnetic flux from the disturbing circuit is:

$$\phi = \int BdS = \int \frac{\mu_0 I}{2\pi r} dS = \frac{\mu_0 I \cdot L}{2\pi} \int_{x_0}^{x_0+x_2} \frac{dr}{r} = \frac{\mu_0 I \cdot L}{2\pi} \ln\left(\frac{x_0 + x_2}{x_0}\right)$$

$$(7.2.8)$$

The induced voltage is:

$$V_2(t) = -N_{TURNS} \frac{\partial \phi}{\partial t} = -1 \cdot \frac{\mu_0 L}{2\pi} \ln\left(\frac{x_0 + x_2}{x_0}\right) \frac{\partial I}{\partial t} =$$

$$= -\mu_0 L I_{MAX} f \ln\left(\frac{x_0 + x_2}{x_0}\right) \cos(2\pi ft)$$

$$(7.2.9)$$

If we write V₂ in fasorial representation:

$$V_2 = \mu_0 L I f \ln\left(\frac{x_0 + x_2}{x_0}\right)$$

$$(7.2.10)$$

In order to calculate the induced current I$_{IND}$ in "C₂" we write:

$$Z_{L2} I_{IND} + Z_{S2} I_{IND} = V_2 \rightarrow I_{IND} = \frac{V_2}{Z_{L2} + Z_{S2}}$$

$$(7.2.11)$$

The voltage applied to the solenoid is:

$$V_S = Z_{L2} I_{IND} = \frac{Z_{L2}}{Z_{L2} + Z_{S2}} V_2$$

$$(7.2.12)$$

If we conservatively neglect the term of Z$_{S2}$ we find:

$$V_S = V_2 = \mu_0 L I f \ln\left(\frac{x_0 + x_2}{x_0}\right)$$

$$(7.2.13)$$

For a typical solenoid valve application:

$I = 10$ A;

$x_0 = 5$ mm;

$x_2 = 3.5$ mm;

$L = 3$ m;

f = 400 Hz.

Using (7.2.13), the maximum induced voltage is V_S = 8.0 mV. For such a solenoid, the minimum energizing voltage is about 4–5 V and so this test does not affect solenoid functioning.

Electric Fields induced into interconnecting cables

This test consists in evaluating the electric fields induced into interconnecting cable by capacitance coupling with another circuit (capacitive alien crosstalk). The test set-up is represented in Figure 7.2-5.

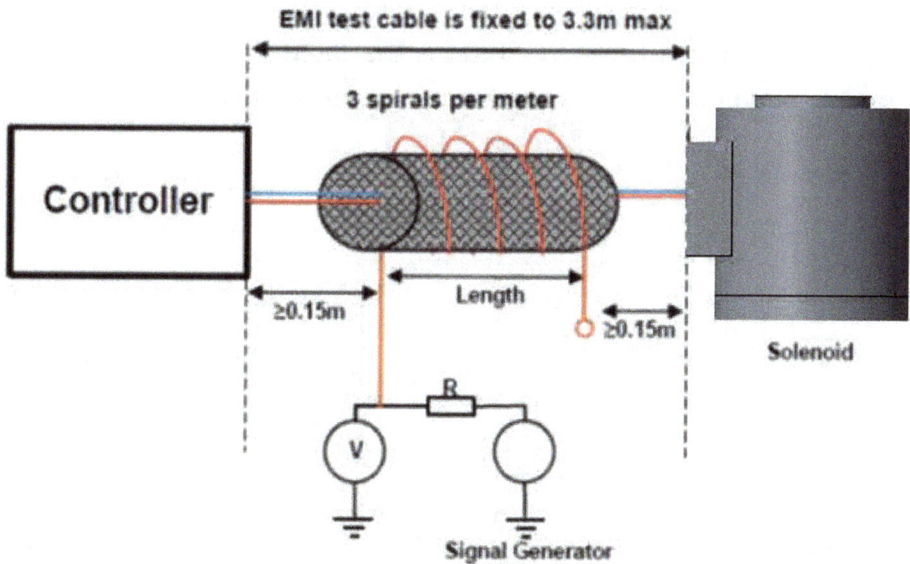

Figure 7.2-5: Electric Fields induced into interconnecting cables: test set up.

We take in account the following conservative assumptions:

- The supply conductors are not twisted;
- No shielding for the supply wires;

Basically we have the situation represented in Figure 7.2-6.

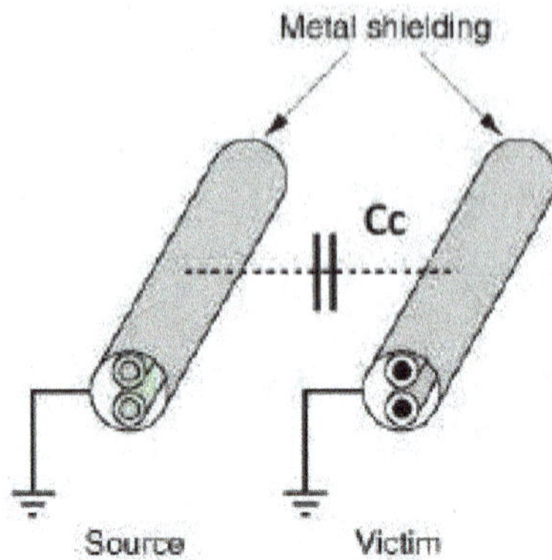

Figure 7.2-6: Electric Fields induced into interconnecting cables: representation.

We have two circuits: the "C_1" circuit is the disturbing one; the "C_2" is the supply circuit. We can represent tha capacitive crosstalk in Figure 7.2-7.

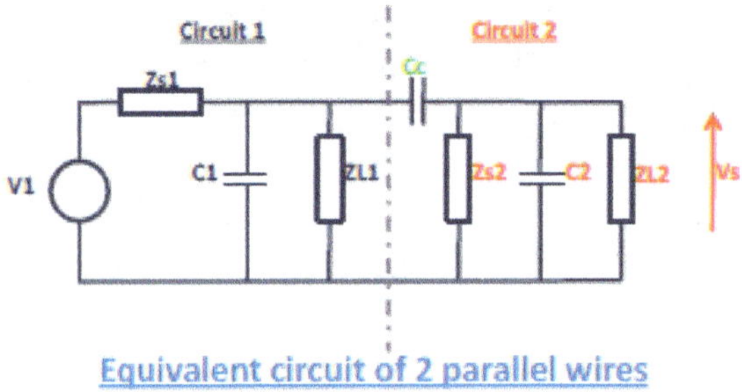

Figure 7.2-7: Electric Fields induced into interconnecting cables: calculation scheme

We are interested to compute the voltage applied on the solenoid V_S. Using a phasor representation, we can write equivalent impedance of "C_2":

$$Z_{EQ} = \frac{1}{\dfrac{1}{Z_{S2}} + \dfrac{1}{Z_{C2}} + \dfrac{1}{Z_{L2}}} \tag{7.2.14}$$

We use an approximation for low frequency:

$$\frac{1}{Z_{C2}} = \omega C_2 \ll 1 \tag{7.2.15}$$

So we can write:

$$Z_{EQ} = \frac{1}{\dfrac{1}{Z_{S2}} + \dfrac{1}{Z_{L2}}} \tag{7.2.16}$$

To simplify computation we assume that $Z_{S2} = Z_{L2} = Z$ (see Figure 7.2-8):

$$Z_{EQ} = \frac{Z}{2}$$
$$Z = \sqrt{R^2 + \omega^2 L^2} \tag{7.2.17}$$

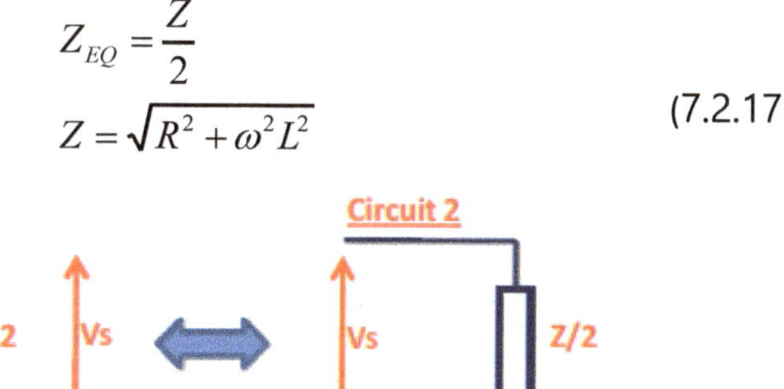

Figure 7.2-8: Electric Fields induced into interconnecting cables: simplification.

In (7.2.17) resistance and inductance of the solenoid appear. We obtain the following equivalent circuit:

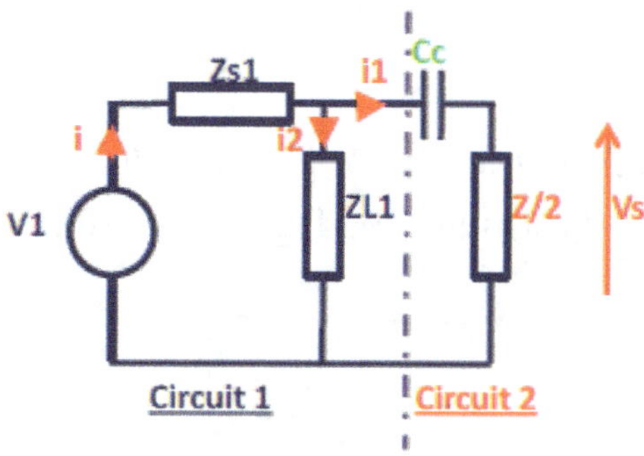

Figure 7.2-9: Electric Fields induced into interconnecting cables: equivalent circuit.

Now we solve this equivalent circuit. The voltage on the solenoid is:

$$V_S = \frac{Z}{2} i_1 \tag{7.2.18}$$

The tension on Z_{L1} is:

$$i_1\left(\frac{1}{C_{CC}\omega} + \frac{Z}{2}\right) = i_2 Z_{L1} \rightarrow i_2 = \frac{i_1}{Z_{L1}}\left(\frac{1}{C_{CC}\omega} + \frac{Z}{2}\right) \tag{7.2.19}$$

The applied voltage on "C_1" is:

$$V_1 = i Z_{S1} + i_2 Z_{L1} = Z_{S1} i_1 + Z_{S1} i_2 + Z_{L1} i_2 = Z_{S1} i_1 + i_2\left(Z_{S1} + Z_{L1}\right) \tag{7.2.20}$$

By neglecting Z_{S1} and substituting (7.2.19) in (7.2.20):

$$V_1 = Z_{S1} i_1 + i_2 Z_{L1} = Z_{S1} i_1 + i_1\left(\frac{1}{C_{CC}\omega} + \frac{Z}{2}\right) = i_1\left(Z_{S1} + \frac{1}{C_{CC}\omega} + \frac{Z}{2}\right) \tag{7.2.21}$$

Expressing i_1 for low frequencies, we have:

$$i_1 = \frac{V_1}{\left(\dfrac{1}{C_{CC}\omega}\right)} \tag{7.2.22}$$

Substituting (7.2.22) into (7.2.18) we obtain:

$$V_S = \frac{Z}{2} C_{CC}\omega V_1 \tag{7.2.23}$$

Now we have to express the C_{CC} for two parallel wires. In literature, we can find the following formula (see Figure 7.2-10):

$$C_{CC} = \frac{\pi \varepsilon_0}{\ln\left[2\left(\dfrac{d}{a}\right)\right]} l \tag{7.2.24}$$

With:

ε_0 = permittivity of air = $1/(36\,\pi\,10^9)$ F/m;

d = 5 mm;

a = 1.37 mm (AWG 22);

l = 3 m.

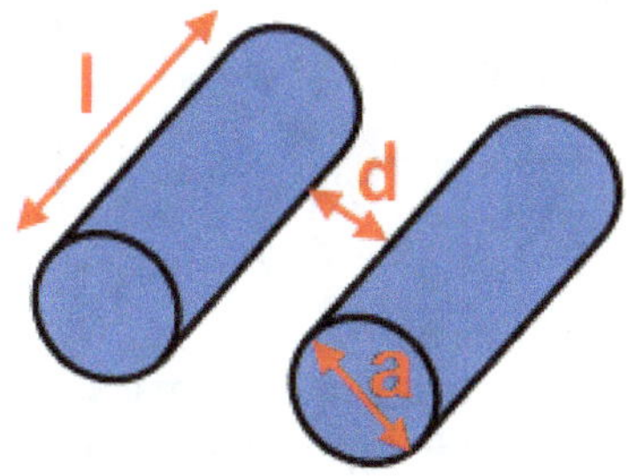

Figure 7.2-10: Electric Fields induced into interconnecting cables: parallel wire capacitance.

We calculate C_{CC} = 42 pF. Therefore, for a typical solenoid with:

R = 80 Ω;

L = 0.6 H;

V_1 = 600 V;

f = 380 Hz.

We can find that V_s = 43.05 mV. For such a solenoid, the minimum energizing voltage is about 4–5 V and so this test does not affect solenoid functioning.

<u>Spike induced into interconnecting cables</u>

These spikes are directly injected on the solenoid pins. They are represented in Figure 7.2-11. The lowest frequency applied on the solenoid is 1/10µs = 100 kHz.

This value is considerably higher than the cut-off frequency of the solenoid (11 Hz). So this test does not affect solenoid functioning.

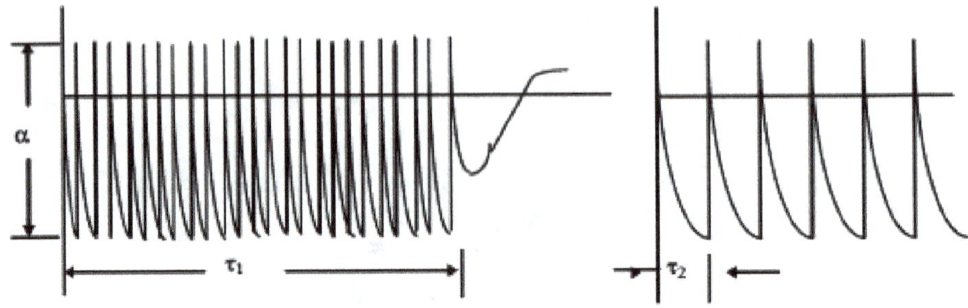

Figure 7.2-11: Spike induced into interconnecting cables: signal type.

7.3. Fly-back diode

Another important protection component is a fly-back diode. This diode is installed in all circuits that are composed by inductive components (as in solenoid valves or actuators). There are many types of diodes but the functioning principle is the same. In Figure 7.3-1 a typical solenoid circuit is shown.

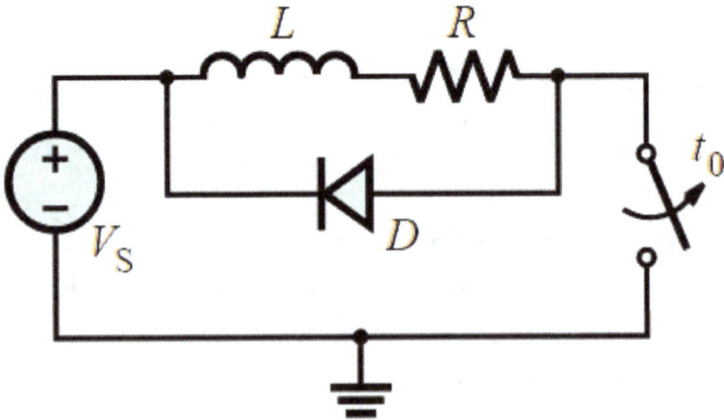

Figure 7.3-1: Typical solenoid circuit with diode.

In Figure 7.3-1, a mono-directional diode is represented. In order to understand the functioning of this component, we have to write the equations of the circuit. For a circuit without diode and opened at t=0 (when $I=I_0$):

$$L\frac{dI(t)}{dt} + R_A I(t) = 0 \qquad (7.3.1)$$

Where R_A is the resistance of the air gap of the opened switch, (so R is negligible). Solving the differential equation:

$$I(t) = I_0 e^{-\frac{R_A}{L}t} \qquad (7.3.2)$$

We can write the voltage across the switch poles as:

$$R_A I(0) = -L\frac{dI(0)}{dt} = I_0 R_A \qquad (7.3.3)$$

This value can be, in typical solenoid applications described here, about 300V. This value will cause an electric arc that can damage the switch or it can be dangerous for fire injection and so on. This phenomenon is called fly-back.

With the presence of diode, the energy stored in the inductor causes an electromotive force that creates a current in the internal loop of resistance, inductance and diode itself (the direction of the current is the one permitted by the diode). So there will not be any voltage across the switch.

Appendix A

Sensitivity analysis

In this section, we will perform a sensitivity analysis on input parameters for solenoid actuator dimensioning. This analysis is based on calculating derivatives of the magnetic force in respect to input parameters, in order to evaluate the changes with respect to current dimensioning configuration. The derivatives will be numerically computed in the current configuration.

We will refer to formula (1.3.6) in chapter 1. The input parameter we will focus on are:

Input parameter	Parameter Description	Current value
D_1	Keeper diameter	11 mm
S_R	Reel thickness	0.9 mm
N_S	Number of radial coils	23
ρ_{MAX}	Maximum current density	16 A/mm^2
ID	Wire identification number	0
ID_{MAT}	Ferromagnetic material identification number	0
ε	Winding efficiency	0.8
D_{STEL}	Stud diameter	4 mm

ID and ID_{MAT} represents respectively the choice of magnet wire and ferromagnetic material. If we change this identification number, we will be able to evaluate the change in magnetic force (in these case we will compute not the derivative but the difference of values). The magnet wires and ferromagnetic materials we considered are:

Magnet wire per IEC 60317			
ID	**D_{IW} (mm)**	**D_W (mm)**	**R_{LIN} (Ω/m)**
0	0.18	0.217	0.6718
1	0.20	0.224	0.5440

Ferromagnetic material		
ID	**μ_r**	**Description**
0	1100	Sandvik 1802®
1	8000	Pure Iron ARMCO®
2	800	AISI 410

Refer to chapter 1 for symbol meaning. Other parameter of formula (1.3.6) that we will not change are:

Fixed parameter	Parameter Description	Current value
T_0	Reference temperature	20°C
T	Ambient temperature	20°C
$V_{e\,NOM}$	Nominal voltage	28 VDC
V_e	Applied voltage	28VDC
l_0	Air gap	0.7 mm
α	Copper temperature coefficient at reference temperature	0.0039 K^{-1}
S_{ISOL}	Insulant thickness	0.15 mm

The magnetic force in the current configuration is:

$$F = 53.387 \text{ N.}$$

Sensitivities are expressed and computed in the following:

$$S_{D_1} = \frac{\partial}{\partial D_1} F\left(D_1, S_R, N_S, \rho_{MAX}, ID, \varepsilon, D_{STEL}, ID_{MAT}\right) = 5.393 \frac{N}{mm}$$

$$S_{S_R} = \frac{\partial}{\partial S_R} F\left(D_1, S_R, N_S, \rho_{MAX}, ID, \varepsilon, D_{STEL}, ID_{MAT}\right) = -11.586 \frac{N}{mm}$$

$$S_{N_S} = \frac{\partial}{\partial N_S} F\left(D_1, S_R, N_S, \rho_{MAX}, ID, \varepsilon, D_{STEL}, ID_{MAT}\right) = -1.007 \frac{N}{turns}$$

$$S_{\rho_{MAX}} = \frac{\partial}{\partial \rho_{MAX}} F\left(D_1, S_R, N_S, \rho_{MAX}, ID, \varepsilon, D_{STEL}, ID_{MAT}\right) = 0.279 \frac{N}{\frac{A}{mm^2}}$$

$$S_{WIRE} = F\left(D_1, S_R, N_S, \rho_{MAX}, ID+1, \varepsilon, D_{STEL}, ID_{MAT}\right) + \\ - F\left(D_1, S_R, N_S, \rho_{MAX}, ID, \varepsilon, D_{STEL}, ID_{MAT}\right) = 26.467 N$$

$$S_{MAT} = F\left(D_1, S_R, N_S, \rho_{MAX}, ID, \varepsilon, D_{STEL}, ID_{MAT}+1\right) + \\ - F\left(D_1, S_R, N_S, \rho_{MAX}, ID, \varepsilon, D_{STEL}, ID_{MAT}\right) = 5.259 N$$

$$S_\varepsilon = \frac{\partial}{\partial \varepsilon} F\left(D_1, S_R, N_S, \rho_{MAX}, ID, \varepsilon, D_{STEL}, ID_{MAT}\right) = 6.141 N$$

$$S_{D_{STEL}} = \frac{\partial}{\partial D_{STEL}} F\left(D_1, S_R, N_S, \rho_{MAX}, ID, \varepsilon, D_{STEL}, ID_{MAT}\right) = -4.068 \frac{N}{mm}$$

If we analyze the results, we can understand that we can act on many parameters in order to set magnetic force to a chosen value. If we want to obtain a solenoid with a force of an higher order of magnitude we need to change magnetic wire, in order to have a thicker wire and so a lower resistance and an higher current absorption.

Appendix B

DC-AC rectifier

This book discusses about DC solenoids. However many solenoid actuators are powered with AC voltage. In order to use DC solenoids in this condition, it is necessary to use a rectifier.

A rectifier is a simple device that converts AC voltage in a DC one. In the following, we will describe the simple full wave rectifier. In Figure B.1 is represented a typical full wave rectifier where we can see the diode bridge, the AC voltage source and a generic load.

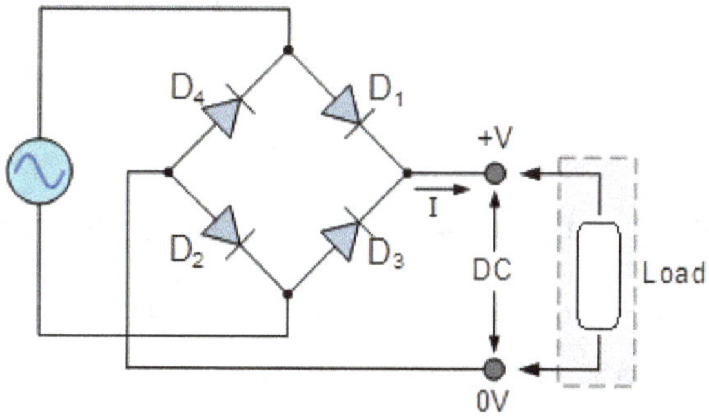

Figure B.1: Full wave rectifier (modified from [25]).

We can analyse the functioning of the rectifier in two different phases. The first one is called positive half-cycle (when source voltage is positive).

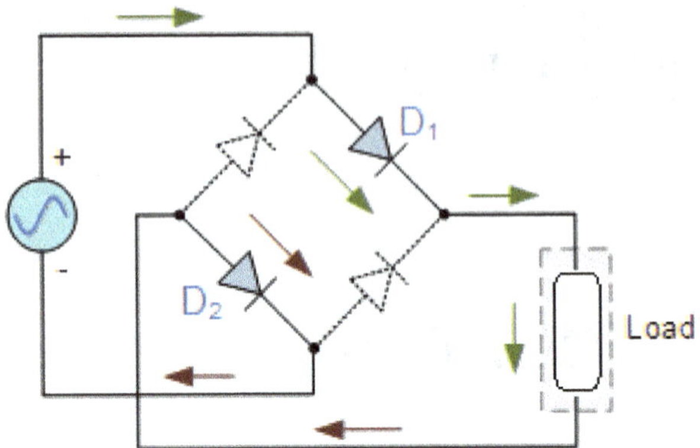

Figure B.2: Positive half-cycle (modified from [25]).

As we can see from Figure B.2, only in diodes D_1 and D_2 there is passage of current. The second phase is called negative half-cycle (when source voltage is negative).

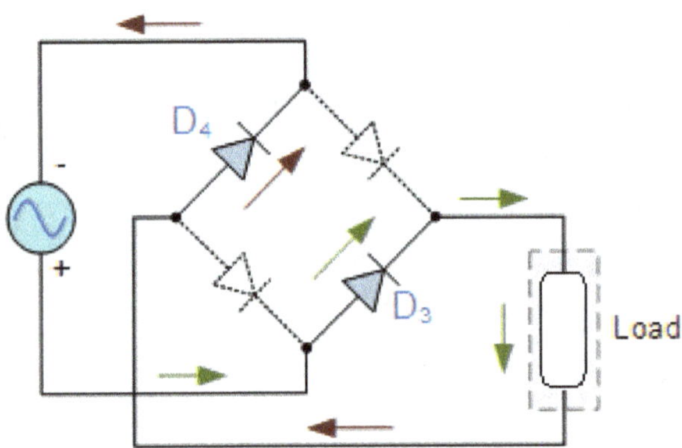

Figure B.3: Negative half-cycle (modified from [25]).

As we can see from Figure B.3, only in diodes D_3 and D_4 there is passage of current. In this way, the voltage on the load is represented in Figure B.4 (curve V_0).

The current on the load depends on the type of load. For resistive load, the current has the same behaviour of the voltage with a value that depends from resistance. Therefore, in this case it is necessary to introduce what is called Smoothing Capacitor (see Figure B.5).

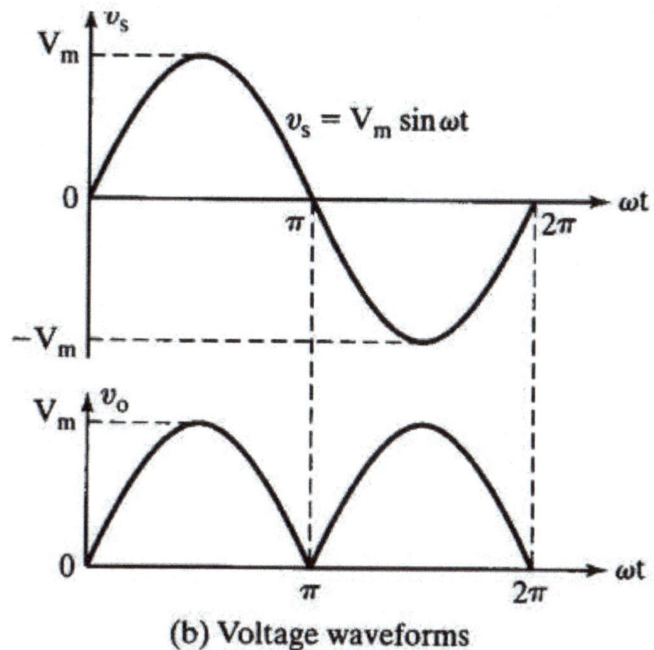

Figure B.4: Source voltage V_s and rectified voltage V_0.

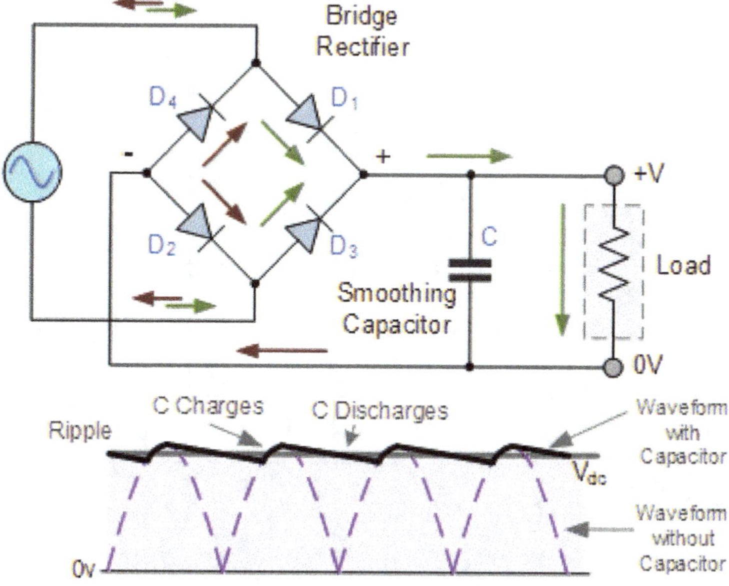

Figure B.5: Full wave rectifier with smoothing capacitor (from [25]).

As we can see in Figure B.5, the charge and discharge of the capacitor gives the possibility to have a nearly direct current. The problem is that capacitors have big dimensions. Moreover, in the case of an inductive load, they are useless. In fact for an inductive load without capacitor, we can solve the circuit by hand and calculate the following results:

$$V(t) = V_m \sin(\omega t)$$
$$V_m = \sqrt{2} V_{RMS}$$

$$i(t) = \frac{V_m}{Z} \left[\sin(\omega t - \theta) + \frac{2\sin(\theta)}{Z\left(1 - e^{-\frac{R\pi}{L\omega}}\right)} e^{-\frac{R}{L}t} \right] \qquad (B.1)$$

$$Z = \sqrt{R^2 + \omega^2 L^2}$$

The result is plotted in Figure B.6 for a solenoid with the following characteristics:

- $R = 35\ \Omega$;
- $L = 0.8\ H$;
- $V_{RMS} = 115\ V_{AC}$

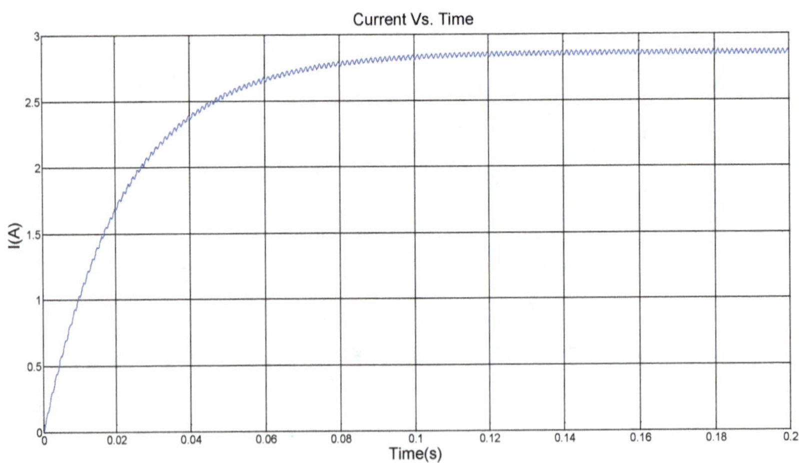

Figure B.6: Current Vs. Time.

We can see that, as in (B.1), current stabilizes on almost steady value, except from little ripples. The steady value is the direct current value and can be calculated also with the following method. The average value of the tension in Figure B.1 is the direct current voltage:

$$V_{DC} = \frac{\int_{0}^{\pi/\omega} V_m \sin(\omega t)\, dt}{\pi/\omega} = \frac{\int_{0}^{\pi} V_m \sin(\theta)\, \frac{d\theta}{\omega}}{\pi/\omega} = \frac{\omega}{\pi}\frac{V_m}{\omega}\int_{0}^{\pi} \sin(\theta)\, d\theta = \frac{2V_m}{\pi} = \frac{2\sqrt{2}V_{RMS}}{\pi}$$

(B.2)

The direct current value:

$$I_{DC} = \frac{V_{DC}}{R} = \frac{2\sqrt{2}V_{RMS}}{\pi R} = 2.96A$$

(B.3)

This is the same value of the average steady value of Figure B.6.

Bibliography

[1] IEC, IEC 60317-13, Polyester or polyesterimide overcoated with polyamide-imide enamelled round copper wire, class 200, Geneve: IEC, 2010.

[2] NEMA, NEMA MW 35-C, Polyester (amide)(imide) Overcoated with Polyamideimide (Single, Heavy, Triple, Quad), Arlington, Virginia: NEMA, 2015.

[3] E. Iurzolla, Il calcolo dei recipienti a pressione, 2a ed., Padova: Edizioni Libreria Cortina, 1981.

[4] J. H. Dellinger, "The temperature coefficient of resistance of copper," *Bullettin of the Bureau of standards*, vol. 7, no. 1, 1911.

[5] J. Takacs, Mathematics of Hysteretic Phenomena, Weinheim: Wiley-VCH, 2006.

[6] AK Steel, «ARMCO® Pure Iron datasheet».

[7] Sandvik Materials Technology, «Sandvik 1802® datasheet».

[8] Metals Handbook, 8th ed., vol. 1, Metals Park, Ohio: American Society of Metals, 1966, p. 794.

[9]   D. De Antonio, "Soft magnetic ferritic stainless steel," *Advanced Materials & Processes,* vol. 161, no. 10, October 2003.

[10] Gruppo Lucefin, «Permeabilità magnetica».

[11] G. Di Caprio, Gli acciai inossidabili, Milano: Hoepli, 2003.

[12] J. R. Brauer, Magnetic actuators and sensors, 1st ed., Hoboken, New Jersey: John Wiley & Sons, Inc., 2006.

[13] M. K. Kazimierczuk e H. Sekiya, «Design of AC resonant inductors using area product method,» in *IEEE Energy Conversion Congress and Exposition*, 2009.

[14] S. Micheletti, «Corso di Metodi Analitici e Numerici per l'Ingegneria Meccanica,» [Online]. Available: http://www1.mate.polimi.it/CN/MANPI0809-Micheletti/191208-Esercitazione/Galerkin-EF1D.pdf.

[15] D. Kuzmin, «Course of Introduction to CFD,» [Online]. Available: http://www.mathematik.uni-dortmund.de/~kuzmin/cfdintro/lecture6.pdf.

[16] K. Stephan and A. Laesecke, "The thermal conductivity of fluid air," *Journal of Phisical and Chemical Reference Data,* vol. 14, no. 1, p. 227, 1985.

[17] Boss, «315 RTV datasheet».

[18] Saint Gobain, «COHRlastic Silicon Rubber datasheet».

[19] D. M. Price e M. Jarratt, «Thermal conductivity of PTFE and PTFE composites,» *Thermochimica Acta,* Vol. %1 di %2392-393, pp. 231-236, 202.

[20] ICAO, ICAO DOC 7488, Manual of the ICAO Standard Atmosphere, Montreal, Quebec: ICAO, 1993.

[21] Metals Handbook, 10th ed., vol. 1, Materials Park, Ohio: ASM International, 1990.

[22] S. Ruben, Handbook of the elements, La Salle, Illinois: Open Court Publishing Company, 1998.

[23] V. Koutsos, «Engineering properties of polymers,» in *ICE Manual of Construction Materials*, vol. II, London, Institution of Civil Engineers, 2009.

[24] RTCA, RTCA-DO-160G, Environmental Conditions and Test Procedures for Airborne Equipment, Washington, DC: RTCA, 2010.

[25] «Electronics Tutorials,» [Online]. Available: http://www.electronics-tutorials.ws/diode/diode_6.html.

[26] L. C. Evans, «Partial Differential Equations,» *Graduate Studies in Mathematics*, vol. 19, 2010.

[27] M. Bertolotti e D. Sette, Lezioni di Fisica: Elettromagnetismo Ottica, Bologna: Zanichelli Editore S.p.A., 2004.

[28] T. Politi, «Corso di Analisi Numerica,» [Online]. Available: http://tiziano19661.interfree.it/pdf1011/Appendice_B.pdf.

Index

Printed in June 2017
by Youcanprint *Self-Publishing*